W9-APR-834

HOLT SCIENCE & TECHNOLOGY

Inside the Restless Earth

HOLT, RINEHART AND WINSTON

A Harcourt Classroom Education Company

Austin • New York • Orlando • Atlanta • San Francisco • Boston • Dallas • Toronto • London

Acknowledgments

Chapter Writers

Kathleen Meehan Berry
Science Chairman
Canon-McMillan School District
Canonsburg, Pennsylvania

Robert H. Fronk, Ph.D.
Chair of Science and Mathematics Education Department
Florida Institute of Technology
West Melbourne, Florida

Mary Kay Hemenway, Ph.D.
Research Associate and Senior Lecturer
Department of Astronomy
The University of Texas
Austin, Texas

Kathleen Kaska
Life and Earth Science Teacher
Lake Travis Middle School
Austin, Texas

Peter E. Malin, Ph.D.
Professor of Geology
Division of Earth and Ocean Sciences
Duke University
Durham, North Carolina

Karen J. Meech, Ph.D.
Associate Astronomer
Institute for Astronomy
University of Hawaii
Honolulu, Hawaii

Robert J. Sager
Chair and Professor of Earth Sciences
Pierce College
Lakewood, Washington

Lab Writers

Kenneth Creese
Science Teacher
White Mountain Junior High School
Rock Springs, Wyoming

Linda A. Culp
Science Teacher and Dept. Chair
Thorndale High School
Thorndale, Texas

Bruce M. Jones
Science Teacher and Dept. Chair
The Blake School
Minneapolis, Minnesota

Shannon Miller
Science and Math Teacher
Llano Junior High School
Llano, Texas

Robert Stephen Ricks
Special Services Teacher
Department of Classroom Improvement
Alabama State Department of Education
Montgomery, Alabama

James J. Secosky
Science Teacher
Bloomfield Central School
Bloomfield, New York

Academic Reviewers

Mead Allison, Ph.D.
Assistant Professor of Oceanography
Texas A&M University
Galveston, Texas

Alissa Arp, Ph.D.
Director and Professor of Environmental Studies
Romberg Tiburon Center
San Francisco State University
Tiburon, California

Paul D. Asimow, Ph.D.
Assistant Professor of Geology and Geochemistry
Department of Physics and Planetary Sciences
California Institute of Technology
Pasadena, California

G. Fritz Benedict, Ph.D.
Senior Research Scientist and Astronomer
McDonald Observatory
The University of Texas
Austin, Texas

Russell M. Brengelman, Ph.D.
Professor of Physics
Morehead State University
Morehead, Kentucky

John A. Brockhaus, Ph.D.
Director—Mapping, Charting, and Geodesy Program
Department of Geography and Environmental Engineering
United States Military Academy
West Point, New York

Michael Brown, Ph.D.
Assistant Professor of Planetary Astronomy
Department of Physics and Astronomy
California Institute of Technology
Pasadena, California

Wesley N. Colley, Ph.D.
Postdoctoral Fellow
Harvard-Smithsonian Center for Astrophysics
Cambridge, Massachusetts

Andrew J. Davis, Ph.D.
Manager—ACE Science Data Center
Physics Department
California Institute of Technology
Pasadena, California

Peter E. Demmin, Ed.D.
Former Science Teacher and Department Chair
Amherst Central High School
Amherst, New York

James Denbow, Ph.D.
Associate Professor
Department of Anthropology
The University of Texas
Austin, Texas

Roy W. Hann, Jr., Ph.D.
Professor of Civil Engineering
Texas A&M University
College Station, Texas

Frederick R. Heck, Ph.D.
Professor of Geology
Ferris State University
Big Rapids, Michigan

Richard Hey, Ph.D.
Professor of Geophysics
Hawaii Institute of Geophysics and Planetology
University of Hawaii
Honolulu, Hawaii

John E. Hoover, Ph.D.
Associate Professor of Biology
Millersville University
Millersville, Pennsylvania

Robert W. Houghton, Ph.D.
Senior Staff Associate
Lamont-Doherty Earth Observatory
Columbia University
Palisades, New York

Steven A. Jennings, Ph.D.
Assistant Professor
Department of Geography & Environmental Studies
University of Colorado
Colorado Springs, Colorado

Eric L. Johnson, Ph.D.
Assistant Professor of Geology
Central Michigan University
Mount Pleasant, Michigan

John Kermond, Ph.D.
Visiting Scientist
NOAA–Office of Global Programs
Silver Spring, Maryland

Zavareh Kothavala, Ph.D.
Postdoctoral Associate Scientist
Department of Geology and Geophysics
Yale University
New Haven, Connecticut

Karen Kwitter, Ph.D.
Ebenezer Fitch Professor of Astronomy
Williams College
Williamstown, Massachusetts

Valerie Lang, Ph.D.
Project Leader of Environmental Programs
The Aerospace Corporation
Los Angeles, California

Philip LaRoe
Professor
Helena College of Technology
Helena, Montana

Julie Lutz, Ph.D.
Astronomy Program
Washington State University
Pullman, Washington

Printed in the United States of America

ISBN 0-03-064784-3

3 4 5 6 7 048 05 04 03 02

Acknowledgments (cont.)

Duane F. Marble, Ph.D.
Professor Emeritus
Department of Geography and Natural Resources
Ohio State University
Columbus, Ohio

Joseph A. McClure, Ph.D.
Associate Professor
Department of Physics
Georgetown University
Washington, D.C.

Frank K. McKinney, Ph.D.
Professor of Geology
Appalachian State University
Boone, North Carolina

Joann Mossa, Ph.D.
Associate Professor of Geography
University of Florida
Gainesville, Florida

LaMoine L. Motz, Ph.D.
Coordinator of Science Education
Department of Learning Services
Oakland County Schools
Waterford, Michigan

Barbara Murck, Ph.D.
Assistant Professor of Earth Science
Erindale College
University of Toronto
Mississauga, Ontario
CANADA

Hilary Clement Olson, Ph.D.
Research Associate
Institute for Geophysics
The University of Texas
Austin, Texas

Andre Potochnik
Geologist
Grand Canyon Field Institute
Flagstaff, Arizona

John R. Reid, Ph.D.
Professor Emeritus
Department of Geology and Geological Engineering
University of North Dakota
Grand Forks, North Dakota

Gary Rottman, Ph.D.
Associate Director
Laboratory for Atmosphere and Space Physics
University of Colorado
Boulder, Colorado

Dork L. Sahagian, Ph.D.
Professor
Institute for the Study of Earth, Oceans, and Space
University of New Hampshire
Durham, New Hampshire

Peter Sheridan, Ph.D.
Professor of Chemistry
Colgate University
Hamilton, New York

David Sprayberry, Ph.D.
Assistant Director for Observing Support
W.M. Keck Observatory
California Association for Research in Astronomy
Kamuela, Hawaii

Lynne Talley, Ph.D.
Professor
Scripps Institution of Oceanography
University of California
La Jolla, California

Glenn Thompson, Ph.D.
Scientist
Geophysical Institute
University of Alaska
Fairbanks, Alaska

Martin VanDyke, Ph.D.
Professor of Chemistry, Emeritus
Front Range Community College
Westminister, Colorado

Thad A. Wasklewicz, Ph.D.
Assistant Professor of Geography
University of Memphis
Memphis, Tennessee

Hans Rudolf Wenk, Ph.D.
Professor of Geology and Geophysical Sciences
University of California
Berkeley, California

Lisa D. White, Ph.D.
Associate Professor of Geosciences
San Francisco State University
San Francisco, California

Lorraine W. Wolf, Ph.D.
Associate Professor of Geology
Auburn University
Auburn, Alabama

Charles A. Wood, Ph.D.
Chairman and Professor of Space Studies
University of North Dakota
Grand Forks, North Dakota

Safety Reviewer

Jack Gerlovich, Ph.D.
Associate Professor
School of Education
Drake University
Des Moines, Iowa

Teacher Reviewers

Barry L. Bishop
Science Teacher and Dept. Chair
San Rafael Junior High School
Ferron, Utah

Yvonne Brannum
Science Teacher and Dept. Chair
Hine Junior High School
Washington, D.C.

Daniel L. Bugenhagen
Science Teacher and Dept. Chair
Yutan Junior & Senior High School
Yutan, Nebraska

Kenneth Creese
Science Teacher
White Mountain Junior High School
Rock Springs, Wyoming

Linda A. Culp
Science Teacher and Dept. Chair
Thorndale High School
Thorndale, Texas

Alonda Droege
Science Teacher
Pioneer Middle School
Steilacom, Washington

Laura Fleet
Science Teacher
Alice B. Landrum Middle School
Ponte Vedra Beach, Florida

Susan Gorman
Science Teacher
Northridge Middle School
North Richland Hills, Texas

C. John Graves
Science Teacher
Monforton Middle School
Bozeman, Montana

Janel Guse
Science Teacher and Dept. Chair
West Central Middle School
Hartford, South Dakota

Gary Habeeb
Science Mentor
Sierra–Plumas Joint Unified School District
Downieville, California

Dennis Hanson
Science Teacher and Dept. Chair
Big Bear Middle School
Big Bear Lake, California

Norman E. Holcomb
Science Teacher
Marion Local Schools
Maria Stein, Ohio

Tracy Jahn
Science Teacher
Berkshire Junior-Senior High School
Canaan, New York

David D. Jones
Science Teacher
Andrew Jackson Middle School
Cross Lanes, West Virginia

Howard A. Knodle
Science Teacher
Belvidere High School
Belvidere, Illinois

Michael E. Kral
Science Teacher
West Hardin Middle School
Cecilia, Kentucky

Kathy LaRoe
Science Teacher
East Valley Middle School
East Helena, Montana

Scott Mandel, Ph.D.
Director and Educational Consultant
Teachers Helping Teachers
Los Angeles, California

Kathy McKee
Science Teacher
Hoyt Middle School
Des Moines, Iowa

Michael Minium
Vice President of Program Development
United States Orienteering Federation
Forest Park, Georgia

Jan Nelson
Science Teacher
East Valley Middle School
East Helena, Montana

Dwight C. Patton
Science Teacher
Carroll T. Welch Middle School
Horizon City, Texas

Joseph Price
Chairman—Science Department
H. M. Brown Junior High School
Washington, D.C.

Terry J. Rakes
Science Teacher
Elmwood Junior High School
Rogers, Arkansas

Steven Ramig
Science Teacher
West Point High School
West Point, Nebraska

Helen P. Schiller
Science Teacher
Northwood Middle School
Taylors, South Carolina

Bert J. Sherwood
Science Teacher
Socorro Middle School
El Paso, Texas

Larry Tackett
Science Teacher and Dept. Chair
Andrew Jackson Middle School
Cross Lanes, West Virginia

Walter Woolbaugh
Science Teacher
Manhattan Junior High School
Manhattan, Montana

Alexis S. Wright
Middle School Science Coordinator
Rye Country Day School
Rye, New York

Gordon Zibelman
Science Teacher
Drexel Hill Middle School
Drexel Hill, Pennsylvania

Inside the Restless Earth

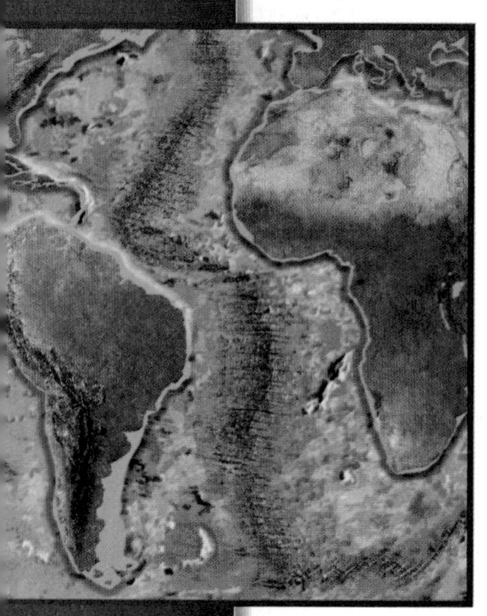

Skills Development

Process Skills

QuickLabs

Chapter Labs

Skills Development

Apply

Feature Articles

Connections

Mathematics

To the Student

This book was created to make your science experience interesting, exciting, and fun!

Go for It!

Science is a process of discovery, a trek into the unknown. The skills you develop using *Holt Science & Technology*—such as observing, experimenting, and explaining observations and ideas—are the skills you will need for the future. There is a universe of exploration and discovery awaiting those who accept the challenges of science.

Science & Technology

You see the interaction between science and technology every day. Science makes technology possible. On the other hand, some of the products of technology, such as computers, are used to make further scientific discoveries. In fact, much of the scientific work that is done today has become so technically complicated and expensive that no one person can do it entirely alone. But make no mistake, the creative ideas for even the most highly technical and expensive scientific work still come from individuals.

Activities and Labs

The activities and labs in this book will allow you to make some basic but important scientific discoveries on your own. You can even do some exploring on your own at home! Here's your chance to use your imagination and curiosity as you investigate your world.

Keep a ScienceLog

In this book, you will be asked to keep a type of journal called a ScienceLog to record your thoughts, observations, experiments, and conclusions. As you develop your ScienceLog, you will see your own ideas taking shape over time. You'll have a written record of how your ideas have changed as you learn about and explore interesting topics in science.

Know "What You'll Do"

The "What You'll Do" list at the beginning of each section is your built-in guide to what you need to learn in each chapter. When you can answer the questions in the Section Review and Chapter Review, you know you are ready for a test.

Check Out the Internet

You will see this sciLINKS NSTA logo throughout the book. You'll be using *sci*LINKS as your gateway to the Internet. Once you log on to *sci*LINKS using your computer's Internet link, type in the *sci*LINKS address. When asked for the keyword code, type in the keyword for that topic. A wealth of resources is now at your disposal to help you learn more about that topic.

In addition to *sci*LINKS you can log on to some other great resources to go with your text. The addresses shown below will take you to the home page of each site.

internet connect

This textbook contains the following on-line resources to help you make the most of your science experience.

Visit **go.hrw.com** for extra help and study aids matched to your textbook. Just type in the keyword HST HOME.

Visit **www.scilinks.org** to find resources specific to topics in your textbook. Keywords appear throughout your book to take you further.

Visit **www.si.edu/hrw** for specifically chosen on-line materials from one of our nation's premier science museums.

Visit **www.cnnfyi.com** for late-breaking news and current events stories selected just for you.

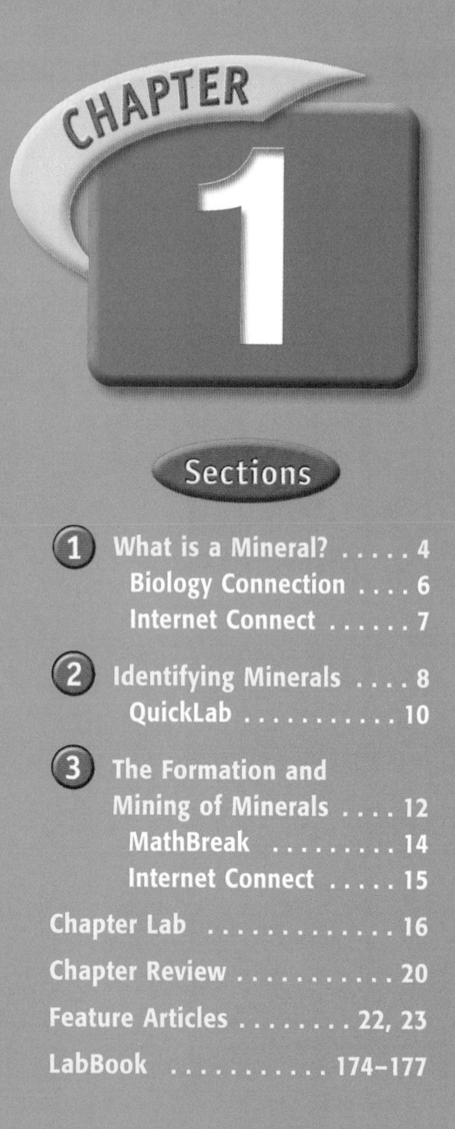

Minerals of the Earth's Crust

Sections

Cave Curtains

Look at the lacy hanging mineral formations in this photo of a limestone cavern. These *stalactites* were formed from dripping water that contained a dissolved mineral, calcium bicarbonate. The mineral reacted with air and hardened into another mineral compound, calcium carbonate. Sometimes called dripstone, such mineral formations may have many colors caused by the presence of yet other minerals in the dripping water. In this chapter, you will learn how mineral compounds form and how they are classified.

1. What is a mineral?
2. How do minerals form?

RIDING A MINERAL?

More than 3,000 different minerals occur naturally on Earth. What is a mineral? Do the following activity, and see if you can figure it out.

Procedure

1. In your ScienceLog, make two columns—one for minerals and one for nonminerals.
2. Ask your classmates what ideas they have about the materials that make up a motorcycle. Take notes as you gather information.
3. Based on what you already know about minerals, classify the materials in a motorcycle into things that come from minerals and things that come from nonminerals.

Analysis

4. Based on your list, is most of a motorcycle made of minerals or nonminerals?
5. Where do you think the minerals that make a motorcycle come from?

SECTION 1

READING WARM-UP

Terms to Learn

mineral
element
atom
compound
crystal
silicate mineral
nonsilicate mineral

What You'll Do

- Explain the four characteristics of a mineral.
- Classify minerals according to the two major compositional groups.

What Is a Mineral?

Not all minerals look like gems. In fact, most of them look more like rocks. But are minerals the same as rocks? Well, not really. So what's the difference? For one thing, rocks are made of minerals, but minerals are not made of rocks. Then what exactly is a mineral? By asking the following four questions, you can tell whether something is a mineral:

A **mineral** is a naturally formed, inorganic solid with a crystalline structure. If you cannot answer "yes" to all four questions above, you don't have a mineral.

How many elements does it take to "set" the periodic table? Find out by turning to page 202.

Minerals: From the Inside Out

Three of the four questions might be easy to answer. The one about crystalline structure may be more difficult. In order to understand what crystalline structure is, you need to know a little about the elements that make up a mineral. **Elements** are pure substances that cannot be broken down into simpler substances by ordinary chemical means. All minerals contain one or more of the 92 elements present in the Earth's crust.

Atoms and Compounds Each element is made of only one kind of atom. An **atom,** as you may recall, is the smallest part of an element that has all the properties of that element. Like all other substances, minerals are made up of atoms of one or more elements.

Most minerals are made of compounds of several different elements. A **compound** is a substance made of two or more elements that have been chemically joined, or bonded together. Halite, for example, is a compound of sodium and chlorine, as shown in **Figure 1.** A few minerals, such as gold and silver, are composed of only one element. For example, pure gold is made up of only one kind of atom—gold.

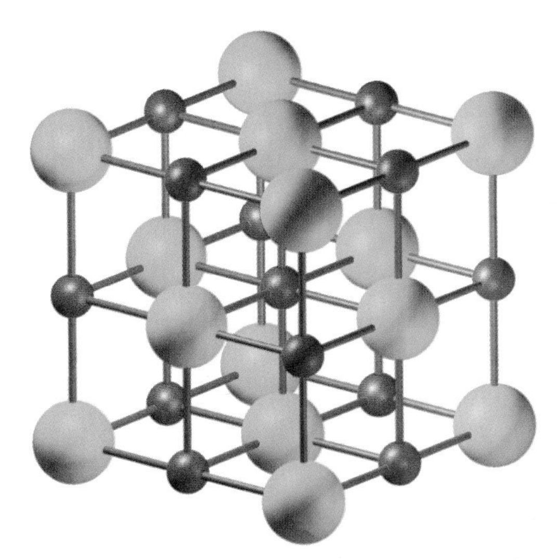

Figure 1 *Atoms of sodium and chlorine are joined together in a compound commonly known as rock salt, or the mineral halite.*

Crystals A mineral is also made up of one or more crystals. **Crystals** are solid, geometric forms of minerals produced by a repeating pattern of atoms that is present throughout the mineral. A crystal's shape is determined by the arrangement of the atoms within the crystal. The arrangement of atoms in turn is determined by the kinds of atoms that make up the mineral. Each mineral has a definite crystalline structure. All minerals can be grouped into crystal classes according to the kinds of crystals they form. **Figure 2** shows how the atomic structure of gold gives rise to cubic crystals.

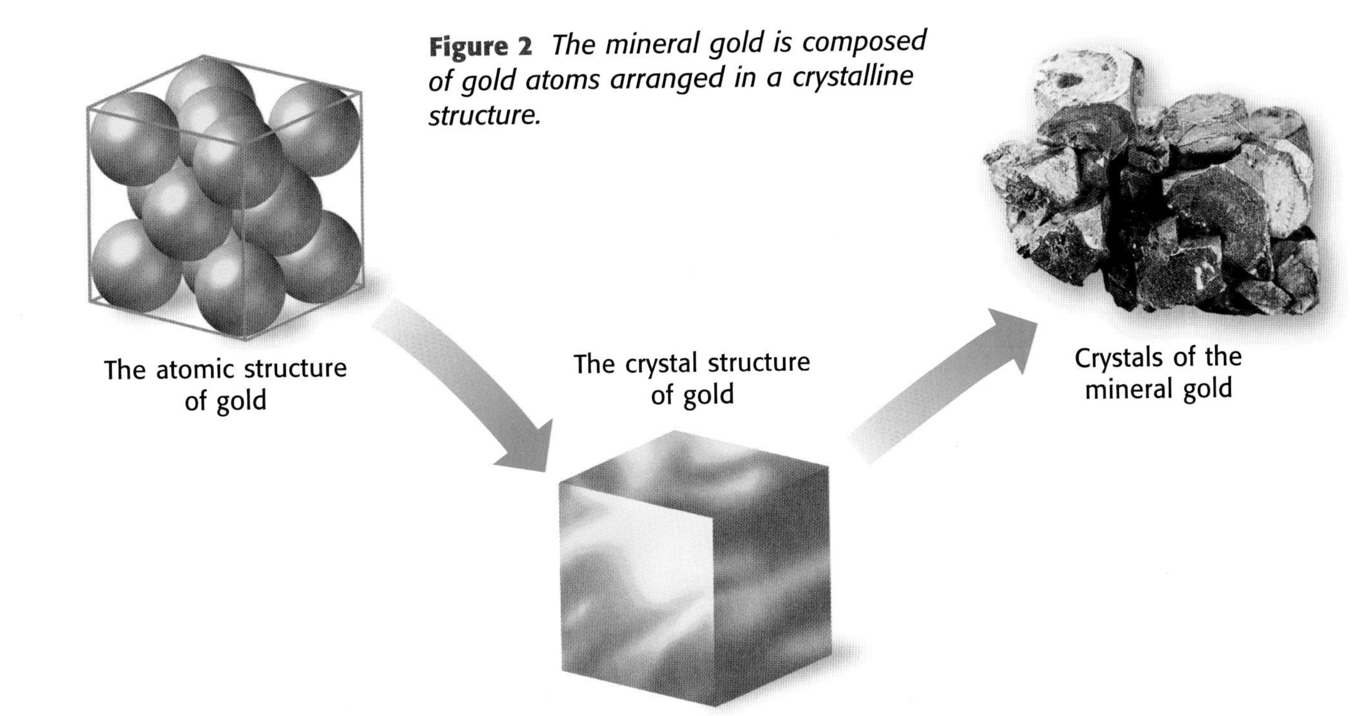

Figure 2 *The mineral gold is composed of gold atoms arranged in a crystalline structure.*

Several species of animals have a brain that contains the mineral magnetite. Magnetite has a special property—it is magnetic. Scientists have shown that certain fish can sense magnetic fields because they have magnetite in their brain. The magnetite gives the fish a sense of direction.

Types of Minerals

Minerals can be classified by a number of different characteristics. The most common classification of minerals is based on chemical composition. Minerals are divided into two groups based on the elements they are composed of. These groups are the silicate minerals and the nonsilicate minerals.

Silicate Minerals Silicon and oxygen are the two most common elements in the Earth's crust. Minerals that contain a combination of these two elements are called **silicate minerals.** Silicate minerals make up more than 90 percent of the Earth's crust—the rest is made up of nonsilicate minerals. Silicon and oxygen usually combine with other elements, such as aluminum, iron, magnesium, and potassium, to make up silicate minerals. Some of the more common silicate minerals are shown in **Figure 3.**

Feldspar Feldspar minerals make up about half the Earth's crust, and they are the main component of most rocks on the Earth's surface. They contain the elements silicon and oxygen along with aluminum, potassium, sodium, and calcium.

Biotite Mica Mica minerals are shiny and soft, and they separate easily into sheets when they break. Biotite is but one of several varieties of mica.

Quartz Quartz (silicon dioxide, SiO_2) is the basic building block of many rocks. If you look closely at the piece of granite, you can see the quartz crystals.

Figure 3 *Granite is a rock composed of various minerals, including feldspar, mica, and quartz.*

Nonsilicate Minerals Minerals that do not contain a combination of the elements silicon and oxygen form a group called the **nonsilicate minerals.** Some of these minerals are made up of elements such as carbon, oxygen, iron, and sulfur. Below are several categories of nonsilicate minerals.

Classes of Nonsilicate Minerals

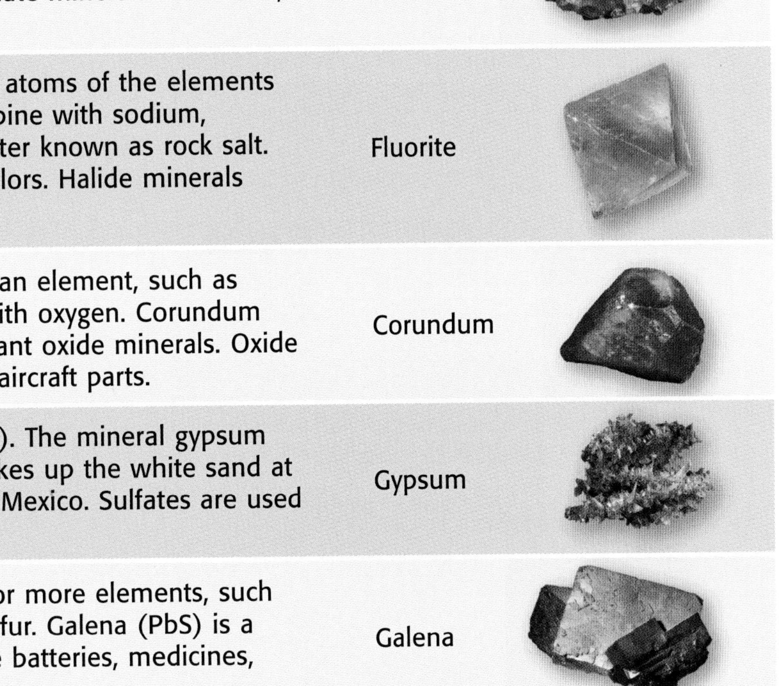

Description	Example
Native elements are minerals that are composed of only one element. About 20 minerals are native elements. Some examples are gold (Au), platinum (Pt), diamond (C), copper (Cu), sulfur (S), and silver (Ag).	Native copper
Carbonates are minerals that contain combinations of carbon and oxygen in their chemical structure. Calcite ($CaCO_3$) is an example of a carbonate mineral. We use carbonate minerals in cement, building stones, and fireworks.	Calcite
Halides are compounds that form when atoms of the elements fluorine, chlorine, iodine, or bromine combine with sodium, potassium, or calcium. Halite (NaCl) is better known as rock salt. Fluorite (CaF_2) can have many different colors. Halide minerals are often used to make fertilizer.	Fluorite
Oxides are compounds that form when an element, such as aluminum or iron, combines chemically with oxygen. Corundum (Al_2O_3) and magnetite (Fe_3O_4) are important oxide minerals. Oxide minerals are used to make abrasives and aircraft parts.	Corundum
Sulfates contain sulfur and oxygen (SO_4). The mineral gypsum ($CaSO_4 \cdot 2H_2O$) is a common sulfate. It makes up the white sand at White Sands National Monument, in New Mexico. Sulfates are used in cosmetics, toothpaste, and paint.	Gypsum
Sulfides are minerals that contain one or more elements, such as lead, iron, or nickel, combined with sulfur. Galena (PbS) is a sulfide. Sulfide minerals are used to make batteries, medicines, and electronic parts.	Galena

SECTION REVIEW

1. What are the differences between atoms, compounds, and minerals?
2. Which two elements are most common in minerals?
3. How are silicate minerals different from nonsilicate minerals?
4. **Making Inferences** Explain why each of the following is not considered a mineral: a cupcake, water, teeth, oxygen.

SECTION 2

READING WARM-UP

Terms to Learn

luster
streak
cleavage
fracture
hardness
density

What You'll Do

- Classify minerals using common mineral-identification techniques.
- Explain special properties of minerals.

Identifying Minerals

If you found the two mineral samples below, how would you know if they were the same mineral?

By looking at these minerals, you can easily see physical similarities. But how can you tell whether they are the same mineral? Moreover, how can you determine the identity of a mineral? In this section you will learn about the different properties that can help you identify minerals.

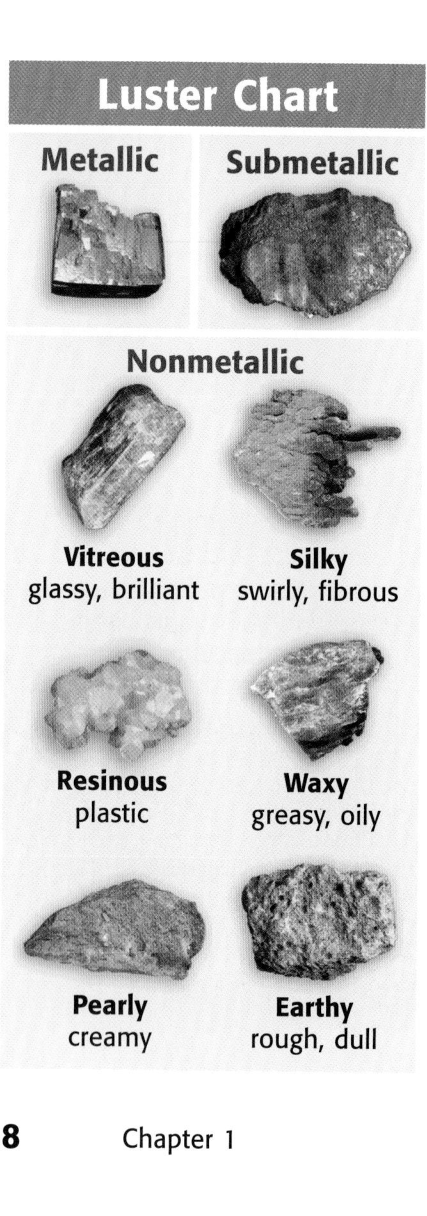

Color

Minerals come in many different colors and shades. The same mineral can come in a variety of colors. For example, in its purest state quartz is clear. Quartz that contains small amounts of impurities, however, can be a variety of colors. Rose quartz gets its color from certain kinds of impurities. Amethyst, another variety of quartz, is purple because it contains other kinds of impurities.

Besides impurities, other factors can change the appearance of minerals. The mineral pyrite, often called fool's gold, normally has a golden color. But if pyrite is exposed to weather for a long period, it turns black. Because of factors such as weathering and impurities, color usually is not a reliable indicator of a mineral's identity.

Luster

The way a surface reflects light is called **luster.** When you say an object is shiny or dull, you are describing its luster. Minerals have metallic, submetallic, or nonmetallic luster. If a mineral is shiny, it may have either a glassy or a metallic luster. If the mineral is dull, its luster is either submetallic or nonmetallic. The different types of lusters are shown in the chart at left.

Streak

The color of a mineral in powdered form is called the mineral's **streak.** To find a mineral's streak, the mineral is rubbed against a piece of unglazed porcelain called a streak plate. The mark left on the streak plate is the streak. The color of a mineral's streak is not always the same as the color of the mineral sample, as shown in **Figure 4.** Unlike the surface of a mineral sample, the streak is not affected by weathering. For this reason, streak is more reliable than color as an indicator of a mineral's identity.

Figure 4 *The color of the mineral hematite may vary, but its streak is always red-brown.*

Cleavage and Fracture

Different types of minerals break in different ways. The way a mineral breaks is determined by the arrangement of its atoms. **Cleavage** is the tendency of some minerals to break along flat surfaces. Gem cutters take advantage of natural cleavage to remove flaws from certain minerals, such as diamonds and rubies, and to shape them into beautiful gemstones. **Figure 5** shows minerals with different cleavage patterns.

Fracture is the tendency of some minerals to break unevenly along curved or irregular surfaces. One type of fracture is shown in **Figure 6.**

Figure 5 *Cleavage varies with mineral type. Mica breaks easily into distinct sheets. Halite breaks at 90° angles in three directions. Diamond breaks in four different directions.*

Figure 6 *This sample of quartz shows a curved fracture pattern called conchoidal (kahn KOYD uhl) fracture.*

QuickLab

Scratch Test

1. You will need a **penny,** a **pencil,** and your **fingernail.** Which one of these three materials is the hardest?
2. Use your fingernail to try to scratch the graphite at the tip of a pencil.
3. Now try to scratch the penny with your fingernail. Which is the hardest of the three?

TRY at HOME

Hardness

Hardness refers to a mineral's resistance to being scratched. If you try to scratch a diamond, you will have a tough time because diamond is the hardest mineral. Talc, on the other hand, is one of the softest minerals. You can scratch it with your fingernail. To determine the hardness of minerals, scientists use *Mohs' hardness scale,* shown below. Notice that talc has a rating of 1 and diamond has a rating of 10. Between these two extremes are other minerals with progressively greater hardness.

To identify a mineral using Mohs' scale, try to scratch the surface of a mineral with the edge of one of the 10 reference minerals. If the reference mineral scratches your mineral, it is harder than your mineral. Continue trying to scratch the mineral until you find a reference mineral that cannot scratch your mineral.

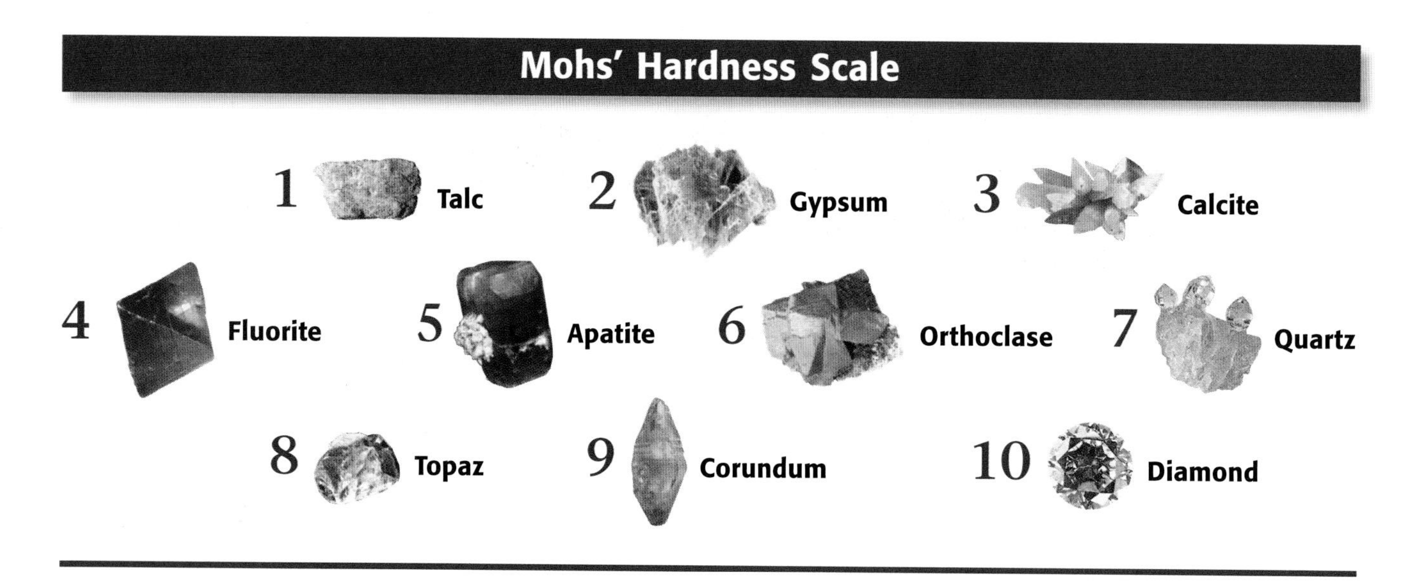

Density

If you pick up a golf ball and a table-tennis ball, which will feel heavier? Although the balls are of similar size, the golf ball will feel heavier because it is denser, as shown in **Figure 7.** **Density** is the measure of how much matter there is in a given amount of space. In other words, density is a ratio of an object's mass to its volume. Density is usually measured in grams per cubic centimeter. Because water has a density of 1 g/cm^3, it is used as a reference point for other substances. The ratio of an object's density to the density of water is called the object's *specific gravity.* The specific gravity of gold, for example, is 19. This means that gold has a density of 19 g/cm^3. In other words, there is 19 times more matter in 1 cm^3 of gold than in 1 cm^3 of water.

Figure 7 *Because a golf ball has a greater density than a table-tennis ball, more table-tennis balls are needed to balance the scale.*

Special Properties

Some properties are particular to only a few types of minerals. The properties below can quickly help you identify the minerals shown. To identify some properties, however, you will need specialized equipment.

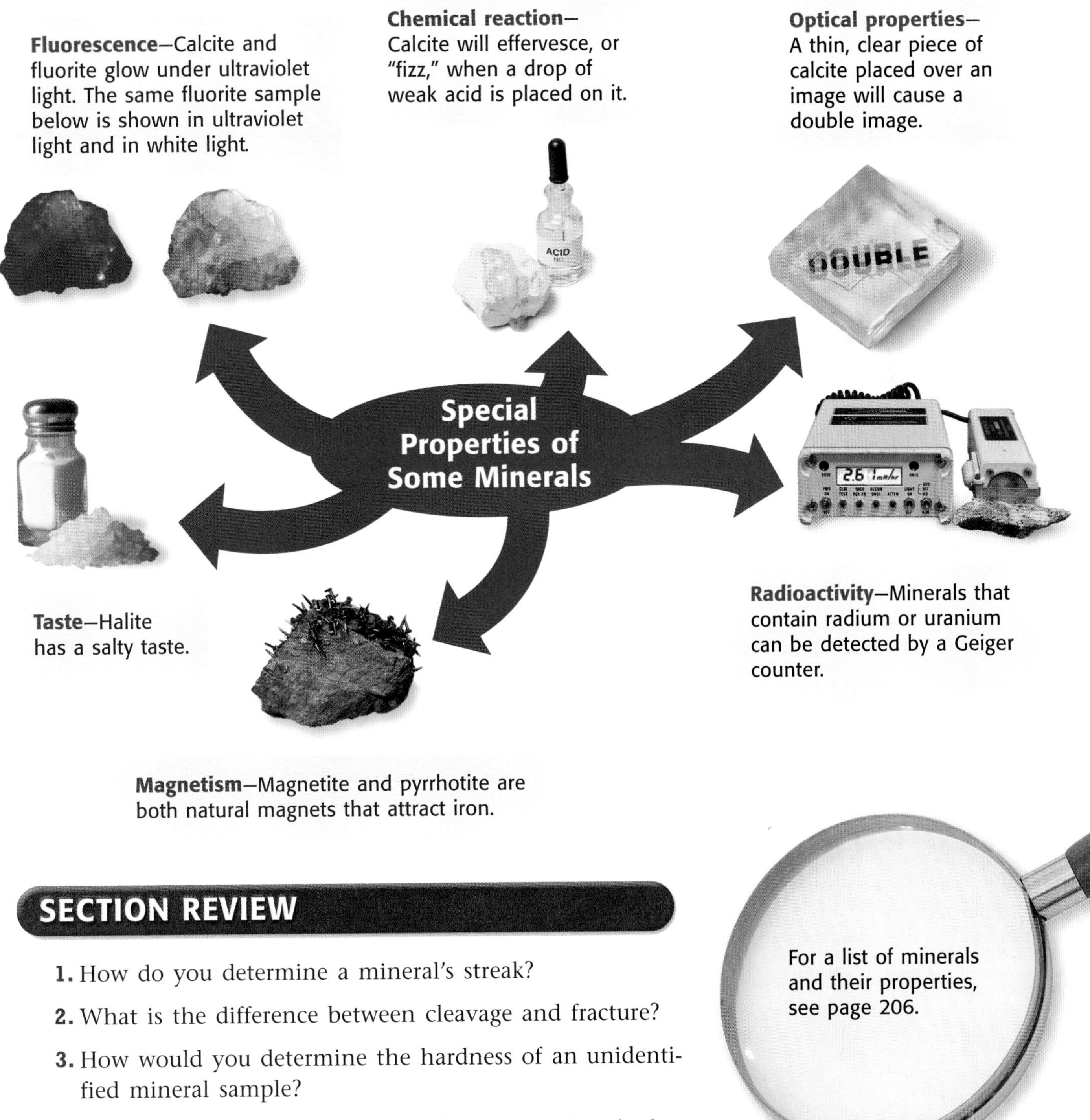

SECTION REVIEW

1. How do you determine a mineral's streak?
2. What is the difference between cleavage and fracture?
3. How would you determine the hardness of an unidentified mineral sample?
4. **Applying Concepts** Suppose you have two minerals that have the same hardness. Which other mineral properties would you use to determine whether the samples are the same mineral?

For a list of minerals and their properties, see page 206.

SECTION 3

READING WARM-UP

Terms to Learn

ore

reclamation

What You'll Do

- Describe the environments in which minerals are formed.
- Compare and contrast the different types of mining.

The Formation and Mining of Minerals

Almost all known minerals can be found in the Earth's crust. They form in a large variety of environments under a variety of physical and chemical conditions. The environment in which a mineral forms determines the mineral's properties. Minerals form both deep beneath the Earth's surface and on or near the Earth's surface.

Evaporating Saltwater When a body of salt water dries up, minerals such as gypsum and halite are left behind. As the salt water evaporates, these minerals crystallize.

Limestones Surface water and ground water carry dissolved materials into lakes and seas, where they crystallize on the bottom. Minerals that form in this environment include calcite and dolomite.

Metamorphic Rocks When changes in pressure, temperature, or chemical makeup alter a rock, *metamorphism* takes place. Minerals that form in metamorphic rock include calcite, garnet, graphite, hematite, magnetite, mica, and talc.

Heat and Pressure

Self-Check

Where do minerals such as gypsum and halite form? *(See page 216 to check your answer.)*

Hot-water Solutions Ground water works its way downward and is heated by magma. It then reacts with minerals to form a hot liquid solution. Dissolved metals and other elements crystallize out of the hot fluid to form new minerals. Gold, copper, sulfur, pyrite, and galena form in such hot-water environments.

Pegmatites As magma moves upward it can form teardrop-shaped bodies called *pegmatites.* The presence of hot fluids causes the mineral crystals to become extremely large, sometimes growing to several meters across! Many gems, such as topaz and tourmaline, form in pegmatites.

Plutons As magma rises upward through the crust, it sometimes stops moving before it reaches the surface and cools slowly, forming millions of mineral crystals. Eventually, the entire magma body solidifies to form a *pluton.* Mica, feldspar, magnetite, and quartz are some of the minerals that form from magma.

Magma

How Pure Is Pure?

Gold classified as 24-karat is 100 percent gold. Gold classified as 18-karat is 18 parts gold and 6 parts another, similar metal. It is therefore 18/24 or 3/4 pure. What is the percentage of pure gold in 18-karat gold?

Mining

Many kinds of rocks and minerals must be mined in order to extract the valuable elements they contain. Geologists use the term **ore** to describe a mineral deposit large enough and pure enough to be mined for a profit. Rocks and minerals are removed from the ground by one of two methods—surface mining or deep mining. The method miners choose depends on how far down in the Earth the mineral is located and how valuable the ore is. The two types of mining are illustrated below.

Surface mining is the removal of minerals or other materials at or near the Earth's surface. Types of surface mines include open pits, strip mines, and quarries. Materials mined in this way include copper ores and bauxite, a mixture of minerals rich in aluminum.

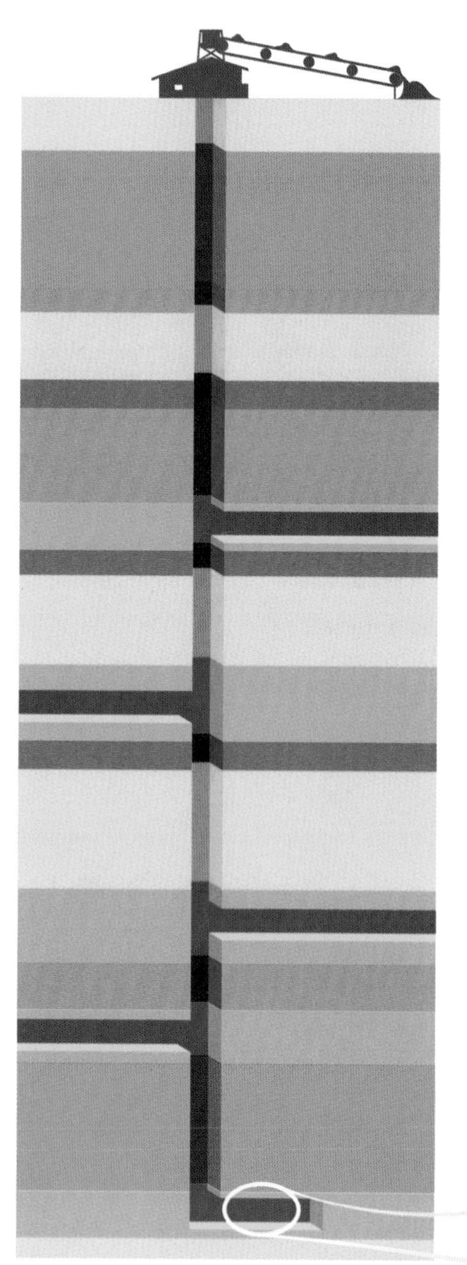

Deep mining is the removal of minerals or other materials from deep within the Earth. Passageways must be dug underground to reach the ore. The retrieval of diamonds and coal commonly requires deep mining.

The Value of Minerals

Many of the metals you are familiar with originally came from mineral ores. You may not be familiar with the minerals, but you will probably recognize the metals extracted from the minerals. The table at right lists some mineral ores and some of the familiar metals that come from them.

As you have seen, some minerals are highly valued for their beauty rather than for their usefulness. Mineral crystals that are attractive and rare are called gems, or gemstones. An example of a gem is shown in **Figure 8.** Gems must be hard enough to be cut and polished.

Common Uses of Minerals

Mineral	Metal	Uses
Chalcopyrite	copper	coins, electrical wire
Galena	lead	batteries, paints
Beryl	beryllium	bicycle frames, airplanes
Chromite	chromium	stainless steel, cast iron, leather tanners

Figure 8 *The Cullinan diamond, at the center of this scepter, is part of the largest diamond ever found.*

Responsible Mining

Mining gives us the minerals we need, but it also creates problems. Mining can destroy or disturb the habitats of plants and animals. The waste products from a mine can get into water sources, polluting both surface water and ground water.

One way to reduce the harmful effects of mining is to return the land to its original state after the mining is completed. This process is called **reclamation.** Reclamation of mined public land has been required by law since the mid-1970s. But reclamation is an expensive and time-consuming process. Another way to reduce the effects of mining is to reduce our need for minerals. We do this by recycling many of the mineral products we currently use, such as aluminum and iron. Mineral ores are *nonrenewable resources;* therefore, the more we recycle, the more we will have in the future.

SECTION REVIEW

1. Describe how minerals form underground.
2. What are the two main types of mining?
3. **Analyzing Ideas** How does reclamation protect the environment around a mine?

internet**connect**

TOPIC: Mining Minerals
GO TO: www.scilinks.org
***sci*LINKS NUMBER:** HSTE070

Skill Builder Lab

Mysterious Minerals

Imagine sitting on a rocky hilltop, gazing at the ground below you. You can see dozens of different types of rocks. How can scientists possibly identify the countless variations? It's a mystery! In this activity you'll use your powers of observation and a few simple tests to determine the identities of rocks and minerals.

MATERIALS

- several sample minerals
- glass microscope slides
- streak plate
- safety gloves
- iron filings

Procedure

1. In your ScienceLog, create a data chart like the one below. Choose one mineral sample. Follow the Mineral Identification Key to find the identity of your sample. When you are finished, record the mineral's name and primary characteristics in the appropriate column in your data chart. **Caution:** Put on your gloves when scratching the glass slide.
2. Select another mineral sample, and repeat the process until your data table is complete.

Mineral Summary Chart

Characteristics	1	2	3	4	5	6
Mineral name						
Luster						
Color						
Streak						
Hardness						
Cleavage						
Special properties						

DO NOT WRITE IN BOOK

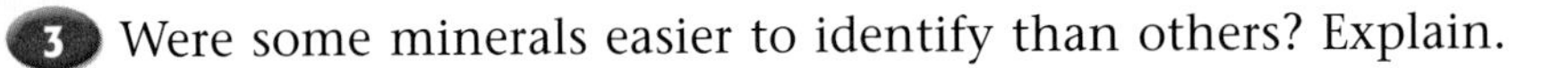

Analysis

3. Were some minerals easier to identify than others? Explain.

4. A streak test is a better indicator of a mineral's true color than visual observation. Why isn't a streak test used to help identify every mineral?

5. In your ScienceLog, summarize what you learned about the various characteristics of each mineral sample you identified.

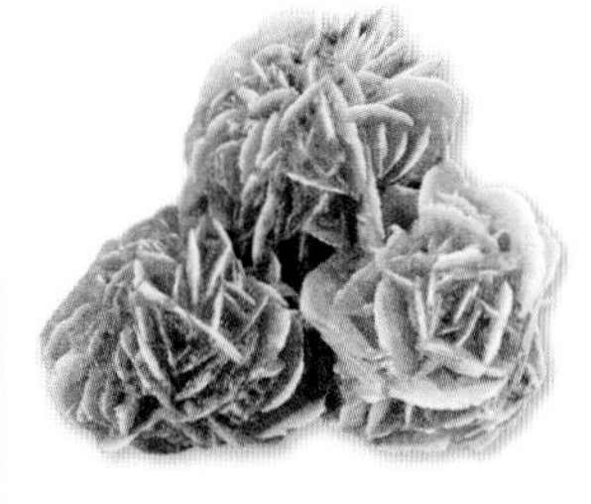

Mineral Identification Key
1. a. If your mineral has a metallic luster, **GO TO STEP 2.** **b.** If your mineral has a nonmetallic luster, **GO TO STEP 3.**
2. a. If your mineral is black, **GO TO STEP 4.** **b.** If your mineral is yellow, it is **PYRITE.** **c.** If your mineral is silver, it is **GALENA.**
3. a. If your mineral is light in color, **GO TO STEP 5.** **b.** If your mineral is dark in color, **GO TO STEP 6.**
4. a. If your mineral leaves a red-brown line on the streak plate, it is **HEMATITE.** **b.** If your mineral leaves a black line on the streak plate, it is **MAGNETITE.** Test your sample for its magnetic properties by holding it near some iron filings.
5. a. If your mineral scratches the glass slide, **GO TO STEP 7.** **b.** If your mineral does not scratch the glass slide, **GO TO STEP 8.**
6. a. If your mineral scratches the glass slide, **GO TO STEP 9.** **b.** If your mineral does not scratch the glass slide, **GO TO STEP 10.**
7. a. If your mineral shows signs of cleavage, it is **ORTHOCLASE FELDSPAR.** **b.** If your mineral does not show signs of cleavage, it is **QUARTZ.**
8. a. If your mineral shows signs of cleavage, it is **MUSCOVITE.** Examine this sample for twin sheets. **b.** If your mineral does not show signs of cleavage, it is **GYPSUM.**
9. a. If your mineral shows signs of cleavage, it is **HORNBLENDE.** **b.** If your mineral does not show signs of cleavage, it is **GARNET.**
10. a. If your mineral shows signs of cleavage, it is **BIOTITE.** Examine this sample for twin sheets. **b.** If your mineral does not show signs of cleavage, it is **GRAPHITE.**

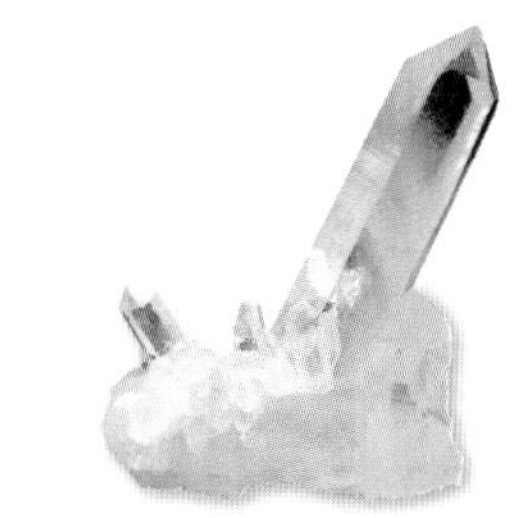

Going Further

Using your textbook and other reference books, research other methods of identifying different types of minerals. Based on your findings, create a new identification key. Give it to a friend along with a few sample minerals, and see if your friend can unravel the mystery!

Chapter Highlights

SECTION 1

Vocabulary

mineral *(p. 4)*
element *(p. 4)*
atom *(p. 5)*
compound *(p. 5)*
crystal *(p. 5)*
silicate mineral *(p. 6)*
nonsilicate mineral *(p. 7)*

Section Notes

- A mineral is a naturally formed, inorganic solid with a definite crystalline structure.
- An atom is the smallest unit of an element that retains the properties of the element.
- A compound forms when atoms of two or more elements bond together chemically.
- Every mineral has a unique crystalline structure. The crystal class a mineral belongs to is directly related to the mineral's chemical composition.
- Minerals are classified as either silicates or nonsilicates. Each group includes different types of minerals.

Labs

Using the Scientific Method *(p. 174)*

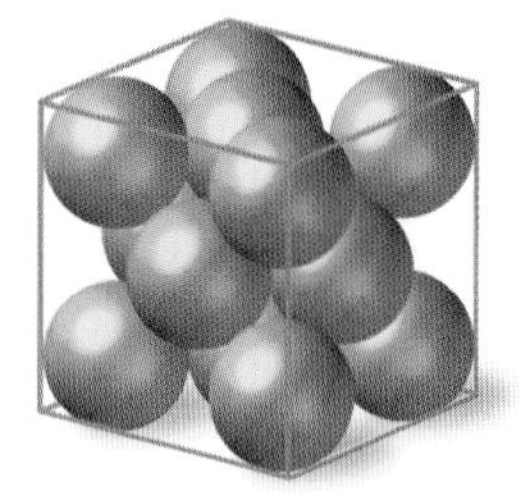

SECTION 2

Vocabulary

luster *(p. 8)*
streak *(p. 9)*
cleavage *(p. 9)*
fracture *(p. 9)*
hardness *(p. 10)*
density *(p. 10)*

Section Notes

- Color is not a reliable indicator for identifying minerals.
- The luster of a mineral can be metallic, submetallic, or nonmetallic.
- A mineral's streak does not necessarily match its surface color.
- The way a mineral breaks can be used to determine its identity. Cleavage and fracture are two ways that minerals break.

Skills Check

Math Concepts

THE PURITY OF GOLD The karat is a measure of the purity of gold. Gold that is 24 karats is 100 percent gold. But gold that is less than 24 karats is mixed with other elements, so it is less than 100 percent gold. If you have a gold nugget that is 16 karats, then 16 parts out of 24 are pure gold—the other 8 parts are composed of other elements.

$$24 \text{ karats} = 100\% \text{ gold}$$
$$16 \text{ karats} = 24 \text{ karats} - 8 \text{ karats}$$
$$\frac{16}{24} = \frac{2}{3} = 0.67 = 67\% \text{ gold}$$

Visual Understanding

ATOMIC STRUCTURE This illustration of the atomic structure of the mineral halite shows that halite is made of two elements—sodium and chlorine. The large spheres represent atoms of chlorine, and the small spheres represent atoms of sodium. The bars between the atoms represent the chemical bonds that hold them together.

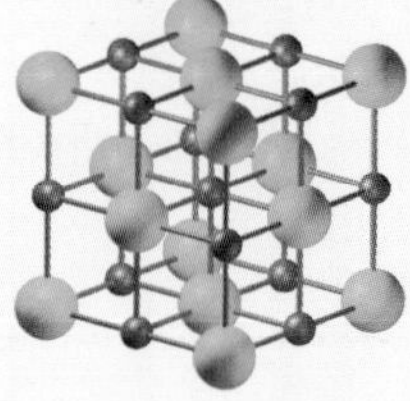

SECTION 2

- Mohs' hardness scale provides a numerical rating for the hardness of minerals.
- The density of a mineral can be used to identify it.
- Some minerals have special properties that can be used to quickly identify them.

Labs

Is It Fool's Gold?—A Dense Situation *(p. 176)*

SECTION 3

Vocabulary

ore *(p. 14)*

reclamation *(p. 15)*

Section Notes

- Minerals form in both underground environments and surface environments.
- Two main types of mining are surface mining and deep mining.
- Minerals are valuable because metals can be extracted from them and because some of them can be cut to form gems.
- Reclamation is the process of returning mined land to its original state.

internet**connect**

GO TO: go.hrw.com

Visit the **HRW** Web site for a variety of learning tools related to this chapter. Just type in the keyword:

KEYWORD: HSTMIN

GO TO: www.scilinks.org

Visit the **National Science Teachers Association** on-line Web site for Internet resources related to this chapter. Just type in the ***sci*LINKS** number for more information about the topic:

TOPIC: Gems — ***sci*LINKS NUMBER:** HSTE055

TOPIC: Birthstones — ***sci*LINKS NUMBER:** HSTE060

TOPIC: Identifying Minerals — ***sci*LINKS NUMBER:** HSTE065

TOPIC: Mining Minerals — ***sci*LINKS NUMBER:** HSTE070

Chapter Review

USING VOCABULARY

For each pair of terms, explain the difference in their meaning.

1. fracture/cleavage
2. element/compound
3. color/streak
4. density/hardness
5. silicate mineral/nonsilicate mineral
6. mineral/atom

UNDERSTANDING CONCEPTS

Multiple Choice

7. On Mohs' hardness scale, which of the following minerals is harder than quartz?
   a. talc **c.** gypsum
   b. apatite **d.** topaz

8. A mineral's streak
   a. is more reliable than color in identifying a mineral.
   b. reveals the mineral's specific gravity.
   c. is the same as a luster test.
   d. reveals the mineral's crystal structure.

9. Which of the following factors is **not** important in the formation of minerals?
   a. heat
   b. volcanic activity
   c. presence of ground water
   d. wind

10. Which of the following terms is **not** used to describe a mineral's luster?
    a. pearly **c.** dull
    b. waxy **d.** hexagonal

11. Which of the following is considered a special property that applies to only a few minerals?
    a. color **c.** streak
    b. luster **d.** magnetism

12. Which of the following physical properties can be expressed in numbers?
    a. luster
    b. hardness
    c. color
    d. reaction to acid

13. Which of the following minerals would scratch fluorite?
    a. talc
    b. quartz
    c. gypsum
    d. calcite

Short Answer

14. Using no more than 25 words, define the term *mineral.*

15. In one sentence, describe how density is used to identify a mineral.

16. What methods of mineral identification are the most reliable? Explain.

Concept Mapping

17. Use the following terms to create a concept map: minerals, oxides, nonsilicates, carbonates, silicates, hematite, calcite, quartz.

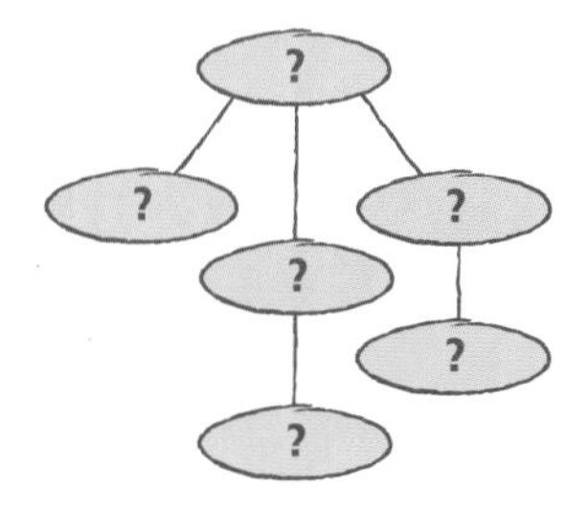

CRITICAL THINKING AND PROBLEM SOLVING

Write one or two sentences to answer the following questions:

18. Suppose you have three rings, each with a different gem. One has a diamond, one has an amethyst (purple quartz), and one has a topaz. You mail the rings in a small box to your friend who lives five states away. When the box arrives at its destination, two of the gems are damaged. One gem, however, is damaged much worse than the other. What scientific reason can you give for the difference in damage?

19. While trying to determine the identity of a mineral, you decide to do a streak test. You rub the mineral across the plate, but it does not leave a streak. Does this mean your test failed? Explain your answer.

20. Imagine that you work at a jeweler's shop and someone brings in some "gold nuggets" that they want to sell. The person claims that an old prospector found the gold nuggets during the California gold rush. You are not sure if the nuggets are real gold. How would you decide whether to buy the nuggets? Which identification tests would help you decide the nuggets' identity?

21. Suppose that you find a mineral crystal that is as tall as you are. What kinds of environmental factors would cause such a crystal to form?

MATH IN SCIENCE

22. Gold has a specific gravity of 19. Pyrite's specific gravity is 5. How much denser is gold than pyrite?

23. In a quartz crystal there is one silicon atom for every two oxygen atoms. That means that the ratio of silicon atoms to oxygen atoms is 1:2. If there were 8 million oxygen atoms in a sample of quartz, how many silicon atoms would there be?

INTERPRETING GRAPHICS

Imagine that you had a sample of feldspar and analyzed it to find out what it is made of. The results of your analysis are shown below.

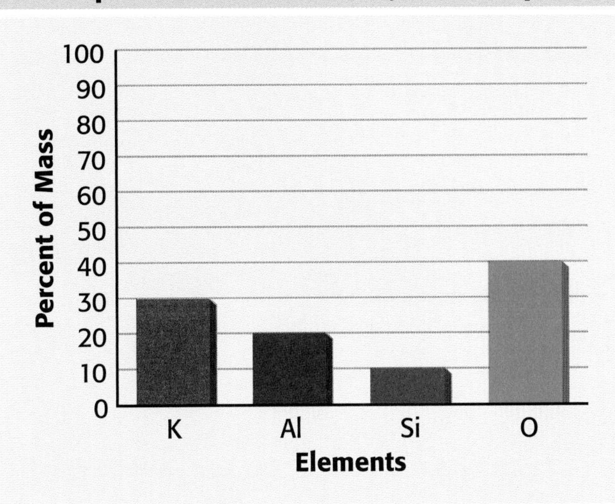

24. Your sample consists of four elements. What percentage of each one is your sample made of?

25. If your mineral sample has a mass of 10 g, how many grams of oxygen does it contain?

26. Make a circle graph showing how much of each of the four elements the feldspar contains. (You will find help on making circle graphs in the Appendix of this book.)

Take a minute to review your answers to the Pre-Reading Questions found at the bottom of page 2. Have your answers changed? If necessary, revise your answers based on what you have learned since you began this chapter.

WEIRD SCIENCE

LIGHTNING LEFTOVERS

Without warning, a bolt of lightning lashes out from a storm cloud and strikes a sandy shoreline with a crash. Almost instantly, the sky is dark again—the lightning has disappeared without a trace. Or has it?

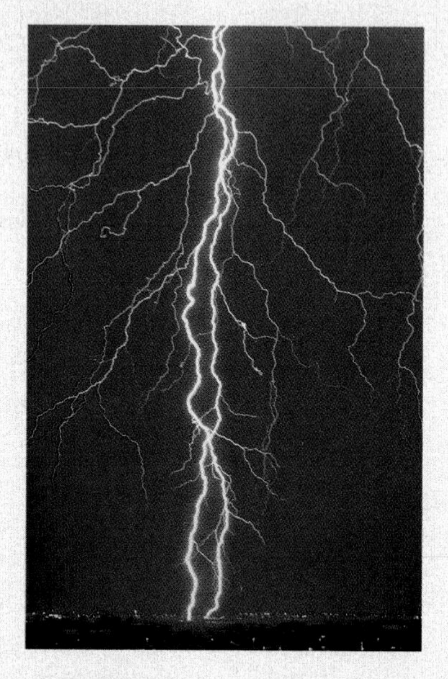

Nature's Glass Factory

Fulgurites are a rare type of natural glass formed when lightning strikes silica-rich minerals that occur commonly in sand, soil, and some rocks. *Tubular fulgurites* are found in areas with a lot of silica, such as beaches or deserts. Lightning creates a tubular fulgurite when a bolt penetrates the sand and melts silica into a liquid. The liquid silica cools and hardens quickly, leaving behind a thin glassy tube, usually with a rough outer surface and a smooth inner surface. Underground, a fulgurite may be shaped like the roots of a tree. It branches out with many arms that trace the zigzag path of the lightning bolt. Some fulgurites are as short as your little finger, while others stretch 20 m into the ground.

Underground Puzzles

So should you expect to run across a fulgurite on your next trip to the beach? Don't count on it. Scientists and collectors search long and hard for the dark glass formations, which often form with little or no surface evidence pointing to their underground location. Even when a fulgurite is located, removing it in one piece is difficult. They are quite delicate, with walls no thicker than 1–2 mm. Some of the largest fulgurites are removed from the ground in many pieces then glued back into their original shape.

Rock Fulgurites

Rock fulgurites are extremely rare, usually occurring only on high mountains. These oddities are created when lightning strikes the surface of a silica-rich rock. A rock fulgurite often looks like a bubbly glass case 1–3 mm thick around the rock. Lightning travels around the outside of the rock, fusing silica-rich minerals on its surface. Depending on which minerals melt, a rock fulgurite's color can range from glassy black to light gray or even bright yellow.

Find Out More

► Investigate how scientists studying the formation of fulgurites try to make lightning bolts strike a precise location to create a new fulgurite. You may also want to do some research to find out about companies that will *create* a fulgurite just for you!

◀ *A Tubular Fulgurite*

Science Fiction

"The Metal Man"

by Jack Williamson

In a dark, dusty corner of Tyburn College Museum stands a life-sized statue of a man. Except for its strange greenish color, the statue looks pretty ordinary. But if you look closely, you will marvel at the perfect detail of the hair and skin. You will also see a strange mark on the statue's chest, a dark crimson shape with six sides.

No one knows how the statue ended up in the dark corner. Everyone believes that the Metal Man is, or once was, Professor Thomas Kelvin of the Geology Department. Professor Kelvin had for many years spent his summer vacations along the Pacific coast of Mexico, prospecting for radium. Then at the end of one summer, Kelvin did not return to Tyburn. He had been more successful than he ever dreamed, and he had become very rich. But high in the mountains, he had also found something else . . .

Now there is only one person who knows what really happened to Professor Kelvin, and he tells the professor's story in "The Metal Man," by Jack Williamson. The tale involves Kelvin's expedition to search for the source of El Rio de la Sangre, the River of Blood, and the radium that makes the river radioactive. Did he find it? Is that what made Kelvin so rich? And what else did Professor Kelvin find there in the remote mountain valley?

Read for yourself the strange story of Professor Kelvin and the Metal Man in the *Holt Anthology of Science Fiction.*

CHAPTER 2

Rocks: Mineral Mixtures

Sections

1. What is the difference between a rock and a mineral?
2. What are some modern uses of rocks?
3. How does rock form?

STEPS FOR A GIANT?

Irish legend claims that the mythical hero Finn MacCool built the Giant's Causeway, shown here. According to legend, these stepping stones were used to cross the sea in order to invade a neighboring island. Actually, this rock formation is the result of the cooling of huge amounts of molten rock. As the molten rock cooled, it formed tall, hexagonal pillars called *columnar joints*. Columnar joints can be seen in basalt rocks around the world. In this chapter, you will learn more about how rocks form.

START-UP Activity

CLASSIFYING OBJECTS

Scientists use the physical and chemical properties of rocks to classify them. Classifying objects such as rocks requires close attention to many properties. Do this exercise to get some classifying practice.

Procedure

1. Your teacher will give you a **bag containing several objects**. Examine the objects and note features such as size, color, shape, texture, smell, and any unique properties.
2. Invent three different ways to sort these objects. You may have only one group or as many as 14.
3. Create an identification key explaining how you organized the objects into each group.

Analysis

4. What properties did you use to sort the items?
5. Were there any objects that could fit into more than one group? How did you solve this problem?
6. Which properties might you use to classify rocks? Explain your answer.

SECTION 1

READING WARM-UP

Terms to Learn

rock	texture
rock cycle	igneous rock
magma	sedimentary rock
composition	metamorphic rock

What You'll Do

- Describe two ways rocks were used by early humans, and describe two ways they are used today.
- Describe how each type of rock changes into another as it moves through the rock cycle.
- List two characteristics of rock that are used to help classify it.

Understanding Rock

The Earth's crust is made up mostly of rock. But what exactly is rock? **Rock** is simply a solid mixture of crystals of one or more minerals. However, some types of rock, such as coal, are made of organic materials. Rocks come in all sizes—from pebbles to formations thousands of kilometers long!

The Value of Rock

Rock has been an important natural resource as long as humans have existed. Early humans used rocks as hammers to make other tools. They discovered that they could make arrowheads, spear points, knives, and scrapers by carefully hammering flint, chert, and obsidian rocks. See **Figure 1.** These rocks were shaped to form extremely sharp edges and points. Even today, obsidian is used to form special scalpels, as shown in **Figure 2.**

Rock has also been used for centuries to make buildings, roads, and monuments. **Figure 3** shows some inventive uses of rock by both ancient and modern civilizations. Buildings have been made out of marble, granite, sandstone, limestone, and slate. Modern buildings also use concrete, in which rock is an important ingredient. Concrete is one of the most common building materials used today.

Figure 1 *This stone tool was made and used more than 5,000 years ago.*

Figure 2 *This stone tool was made recently. It is an obsidian scalpel used in delicate operations.*

Figure 3 *These photos show a few samples of structures built with rock. On this page are structures built by ancient civilizations. On the facing page are some more-modern examples.*

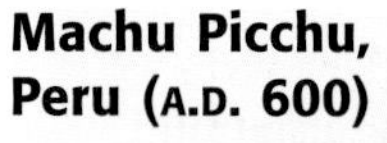

Pyramids at Giza, Egypt (3000 B.C.)

Machu Picchu, Peru (A.D. 600)

Humans have a long history with rock. Certain types of rock have helped us to survive and to develop both our ancient and modern civilizations. Rock is also very important to scientists. The study of rocks helps answer questions about the history of the Earth and our solar system. Rocks provide a record of what the Earth and other planets were like before recorded history.

The fossils some rocks contain also provide clues about lifeforms that lived billions of years ago, long before dinosaurs walked the Earth. **Figure 4** shows how rocks can capture evidence of life that became extinct long ago. Without such fossils, scientists would know very little about the history of life on Earth. The answers we get from studying rocks often cause us to ask even more questions!

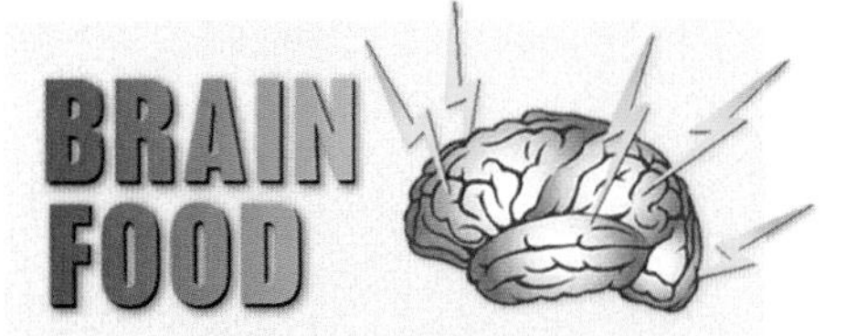

Some meteorites are actually rocks that come from other planets. Below is a microscopic view of a meteorite that came from Mars. The tiny structures may indicate that microscopic life once existed on Mars.

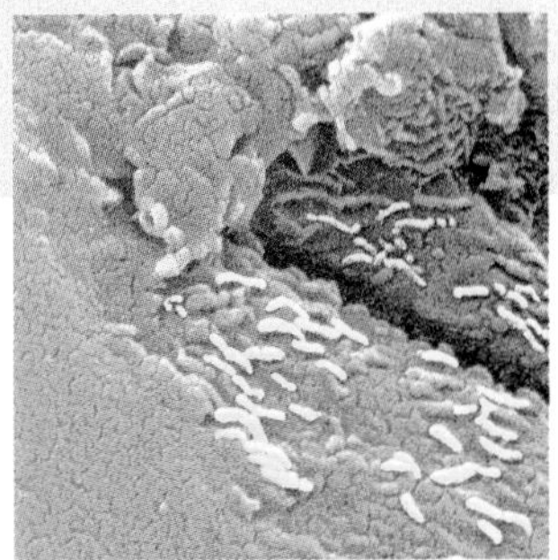

Figure 4 *These fossils were found on a mountaintop. Their presence indicates that what is now a mountaintop was once the bottom of a shallow sea.*

Exeter Cathedral, Exeter, England (A.D. 1120–1520)

LBJ Library, Austin, Texas (1972)

The Rock Cycle

The rocks in the Earth's crust are constantly changing. Rock changes its shape and composition in a variety of ways. The way rock forms determines what type of rock it is. The three main types of rock are *igneous, sedimentary,* and *metamorphic.* Each type of rock is a part of the *rock cycle.* The **rock cycle** is the process by which one rock type changes into another. Follow this diagram to see one way sand grains can change as they travel through the rock cycle.

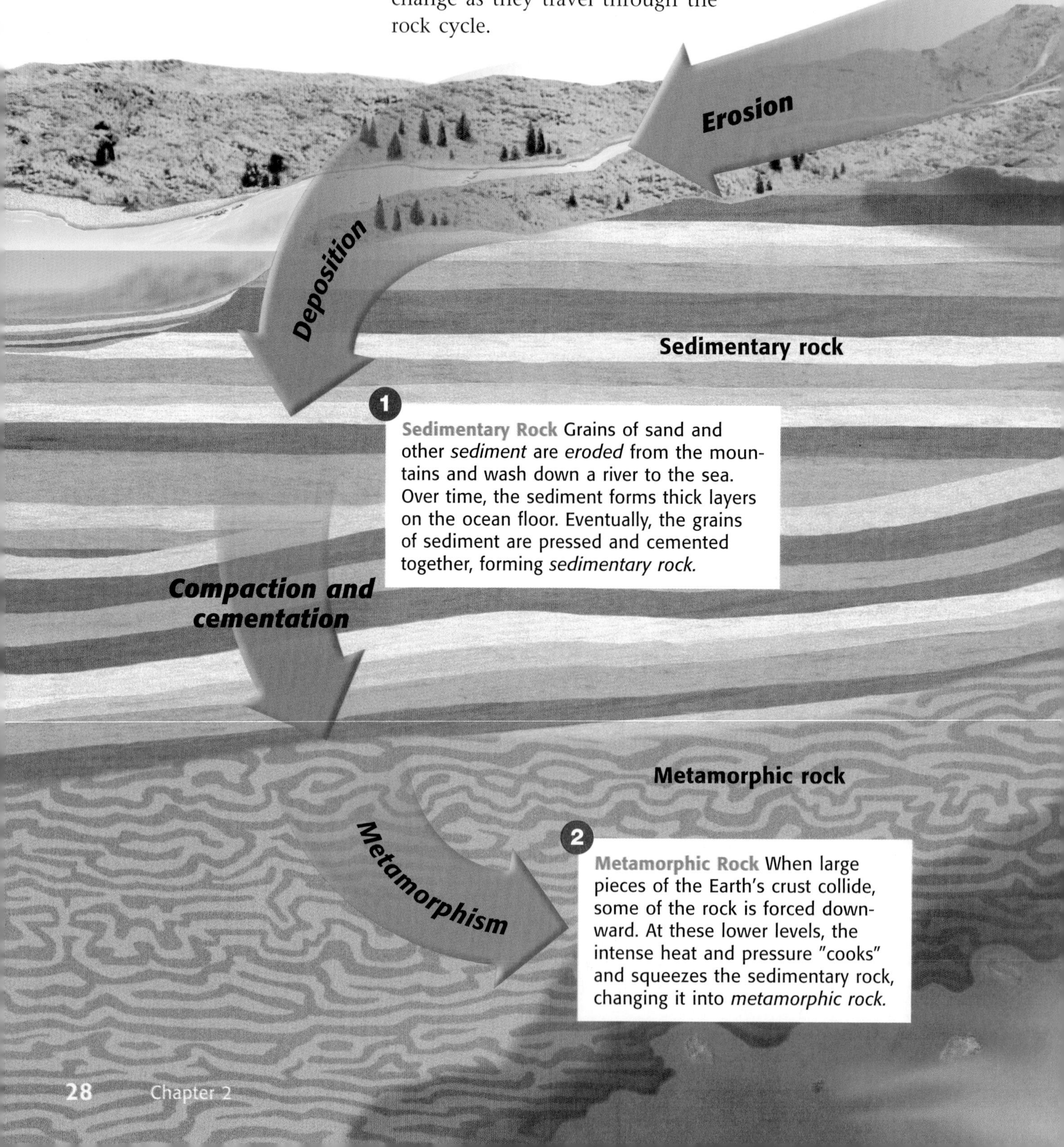

Erosion

Deposition

Sedimentary rock

1 **Sedimentary Rock** Grains of sand and other *sediment* are *eroded* from the mountains and wash down a river to the sea. Over time, the sediment forms thick layers on the ocean floor. Eventually, the grains of sediment are pressed and cemented together, forming *sedimentary rock.*

Compaction and cementation

Metamorphic rock

Metamorphism

2 **Metamorphic Rock** When large pieces of the Earth's crust collide, some of the rock is forced downward. At these lower levels, the intense heat and pressure "cooks" and squeezes the sedimentary rock, changing it into *metamorphic rock.*

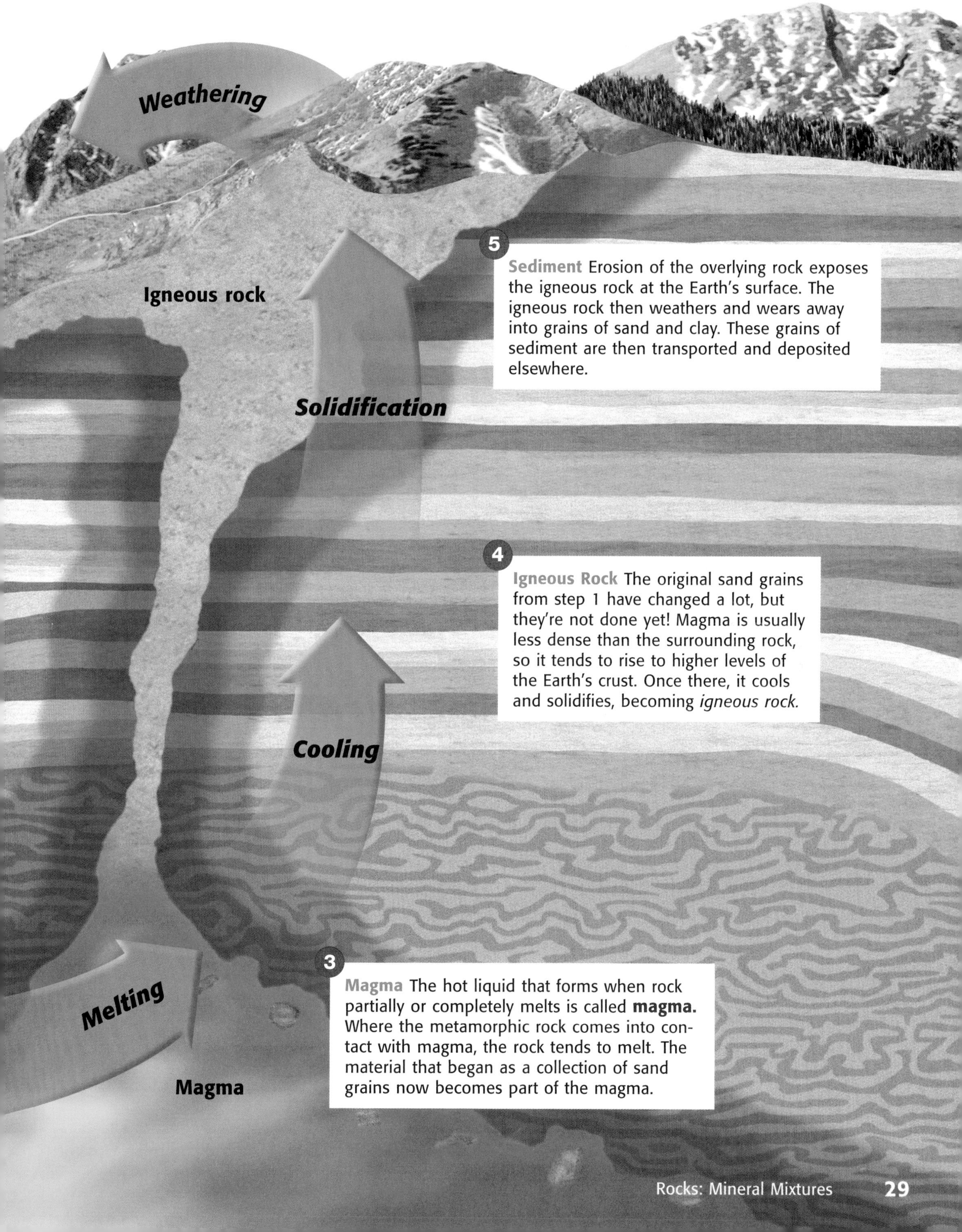

Weathering

Igneous rock

5

Sediment Erosion of the overlying rock exposes the igneous rock at the Earth's surface. The igneous rock then weathers and wears away into grains of sand and clay. These grains of sediment are then transported and deposited elsewhere.

Solidification

4

Igneous Rock The original sand grains from step 1 have changed a lot, but they're not done yet! Magma is usually less dense than the surrounding rock, so it tends to rise to higher levels of the Earth's crust. Once there, it cools and solidifies, becoming *igneous rock.*

Cooling

3

Magma The hot liquid that forms when rock partially or completely melts is called **magma.** Where the metamorphic rock comes into contact with magma, the rock tends to melt. The material that began as a collection of sand grains now becomes part of the magma.

Melting

Magma

Now that you know something about the natural processes that make the three major rock types, you can see that each type of rock can become any other type of rock. This is why it is called a cycle—there is no beginning or end. All rocks are at some stage of the rock cycle and can change into a different rock type. **Figure 5** shows how the three types of rock change form.

Figure 5 The Rock Cycle

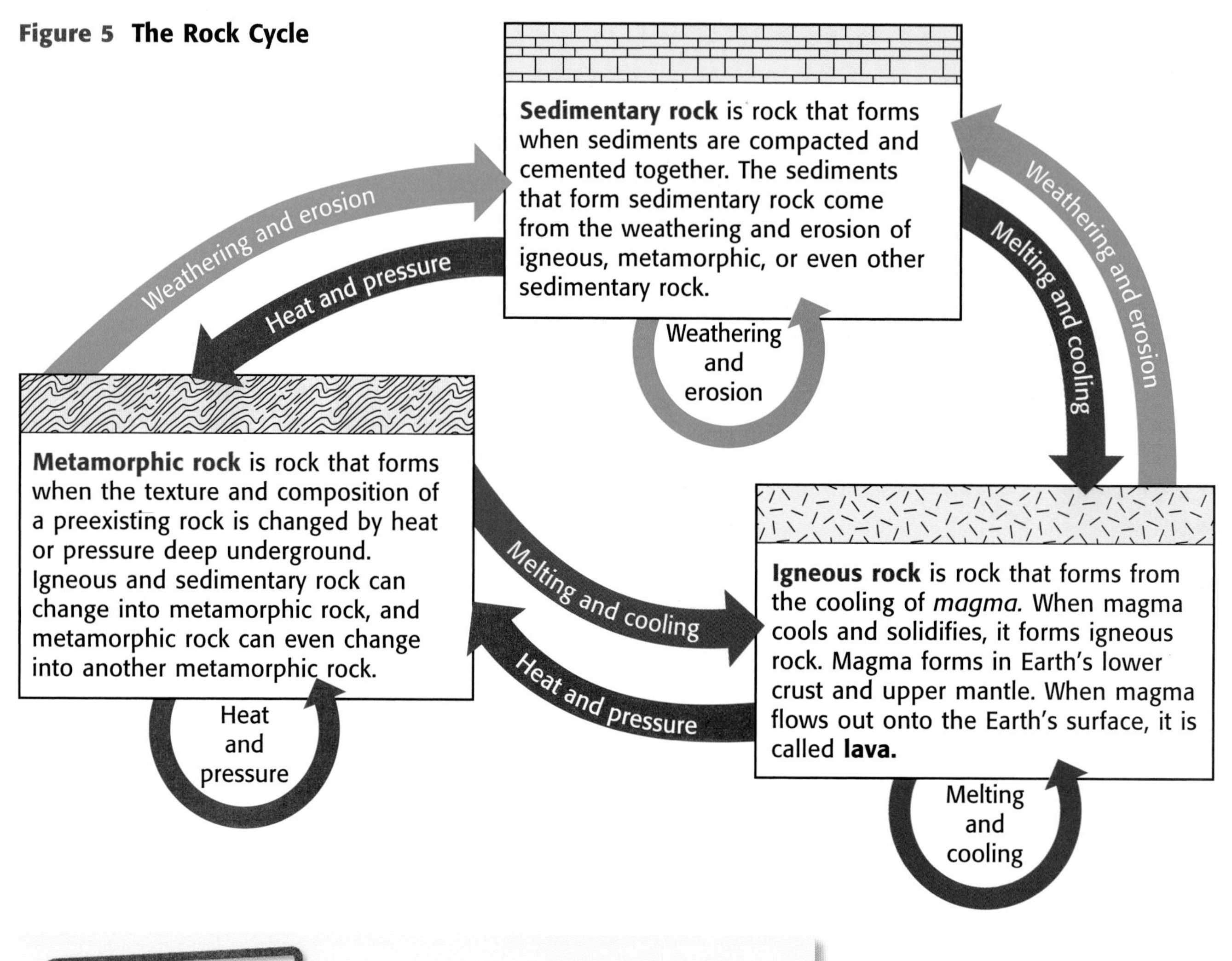

Classifying Objects

Geologists sometimes use food examples to describe geologic processes. For example, when you tap a block of gelatin, it shakes much like the ground shakes during an earthquake. Melting and solidifying chocolate chips models the formation of magma and igneous rocks. Think of a way that food can be used to describe the formation of sedimentary and metamorphic rocks. What do you think are the strengths and weaknesses of using food to describe geologic processes? Explain.

The Nitty-Gritty on Rock Classification

You now know that scientists classify all rock into three main types based on how they formed. But did you know that each type of rock is divided into even smaller groups? These smaller groups are also based on differences in the way rocks form. For example, all igneous rock forms when hot liquid cools and solidifies. But some igneous rocks form when lava cools on the Earth's surface, while others form when magma cools deep beneath the surface. Therefore, igneous rock is divided into two smaller groups, depending on how and where it forms. In the same way, sedimentary and metamorphic rocks are also divided into smaller groups. How do Earth scientists know how to classify different rocks? They study them in detail using two important criteria—*composition* and *texture.*

What's in It?

Assume that a granite rock you are studying is made of 30 percent quartz, 55 percent feldspar, and the rest biotite mica. What percentage of the rock is biotite mica?

Composition The minerals a rock is made of determine the **composition** of the rock. For example, a rock that is made up mostly of the mineral quartz will have a composition very similar to quartz. A rock that is made of 50 percent quartz and 50 percent feldspar will have a very different overall composition. Use this idea to compare the examples given in **Figure 6.**

Figure 6 *The overall composition of a rock depends on the minerals it contains.*

Texture The **texture** of a rock is determined by the sizes, shapes, and positions of the grains of which it is made. Rocks that are made entirely of small grains, such as silt or clay particles, are said to have a *fine-grained* texture. Rocks that are made of large grains, such as pebbles, are said to have a *coarse-grained* texture. Rocks that have a texture between fine- and coarse-grained are said to have a *medium-grained* texture. Examples of these textures are shown in **Figure 7.**

Figure 7 *These three sedimentary rocks are made up of grains of different sizes. Can you see the differences in their textures?*

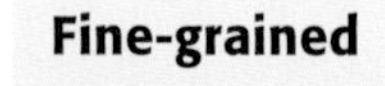

Siltstone

Medium-grained

Sandstone

Coarse-grained

Conglomerate

Each rock type has a different kind of texture that can provide good clues to how and where the rock formed. For example, the rock shown in **Figure 8** has a texture that reflects how it formed. Both texture and composition are important characteristics that scientists use to understand the origin and history of rocks. Keep these characteristics in mind as you continue reading through this chapter.

Figure 8 *This layered sandstone formed at the bottom of a river. The sediments from which it is made were deposited in layers.*

internet**connect**

SCILINKS NSTA

TOPIC: Composition of Rock
GO TO: www.scilinks.org
***sci*Links Number:** HSTE090

SECTION REVIEW

1. List two ways rock is important to humans today.
2. What are the three major rock types, and how can they change from one type to another type?
3. How is lava different from magma?
4. **Comparing Concepts** Explain the difference between texture and composition.

SECTION 2

READING WARM-UP

Terms to Learn

intrusive
extrusive

What You'll Do

- Explain how the cooling rate of magma affects the properties of igneous rock.
- Distinguish between igneous rock that cools deep within the crust and igneous rock that cools at the surface.
- Identify common igneous rock formations.

Igneous Rock

The word *igneous* comes from the Latin word for "fire." Magma cools into various types of igneous rock depending on the composition of the magma and the amount of time it takes the magma to cool and solidify. Like all other rock, igneous rock is classified according to its composition and texture.

Origins of Igneous Rock

Magma and lava solidify in much the same way that water freezes. When magma or lava cools down enough, it solidifies, or "freezes," to form igneous rock. One difference between water freezing and magma freezing is that water freezes at 0°C and magma and lava freeze at between 700°C and 1,250°C.

There are three ways magma can form: when rock is heated, when pressure is released, or when rock changes composition. To see how this can happen, follow along with **Figure 9.**

Figure 9 *There are three ways a rock can melt.*

Temperature An increase in temperature deep within the Earth's crust can cause the minerals in a rock to melt. Different minerals melt at different temperatures. So depending on how hot a rock gets, some of the minerals can melt while other minerals remain solid.

Pressure The high pressure deep within the Earth forces minerals to stay in the solid state, when otherwise they would melt from the intense heat. When hot rocks rise to shallow depths, the pressure is finally released and the minerals can melt.

Composition Sometimes fluids like water and carbon dioxide enter a rock that is close to its melting point. When these fluids combine with the rock, they can lower the melting point of the rock enough for it to melt and form magma.

Composition and Texture of Igneous Rock

Look at the rocks in **Figure 10.** All of these are igneous rocks, even though they look very different from one another. These rocks differ from one another in what they are made of and how fast they cooled.

The light-colored rocks are not only lighter in color but also less dense. They are rich in elements such as silicon, aluminum, sodium, and potassium. These lightweight rocks are called *felsic.* The darker rocks are denser than the felsic rocks. These rocks are rich in iron, magnesium, and calcium and are called *mafic.*

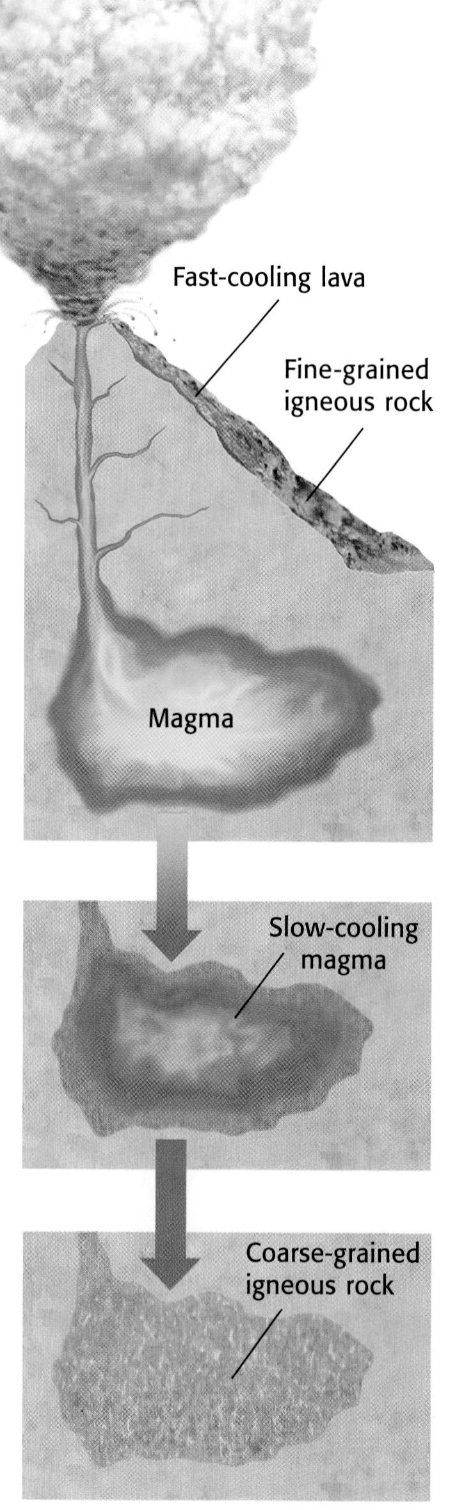

Figure 11 *The amount of time it takes for magma or lava to cool determines the texture of igneous rock.*

Figure 10 *Light-colored igneous rock generally has a felsic composition. Dark-colored igneous rock generally has a mafic composition.*

Now look at **Figure 11.** This illustration shows what happens to magma when it cools at different rates. The longer it takes for the magma or lava to cool, the more time mineral crystals have to grow. And the more time the crystals have to grow, the coarser the texture of the resulting igneous rock.

Self-Check

Rank the rocks shown in Figure 10 by how fast they cooled. Hint: Pay attention to their texture. *(See page 216 to check your answer.)*

Igneous Rock Formations

You have probably seen igneous rock formations that were caused by lava cooling on the Earth's surface. But not all magma reaches the surface. Some magma cools and solidifies deep within the Earth's crust.

Intrusive Igneous Rock When magma cools beneath the Earth's surface, the resulting rock is called **intrusive.** Intrusive rock usually has a coarse-grained texture. This is because it is well insulated by the surrounding rock and thus cools very slowly.

Intrusive rock formations are named for their size and the way in which they intrude, or push into, the surrounding rock. *Plutons* are large, balloon-shaped intrusive formations that result when magma cools at great depths. **Figure 12** shows an example of an intrusive formation that has been exposed on the Earth's surface. Some common intrusive rock formations are shown in **Figure 13.**

Figure 12 *Enchanted Rock, near Llano, Texas, is an exposed pluton made of granite.*

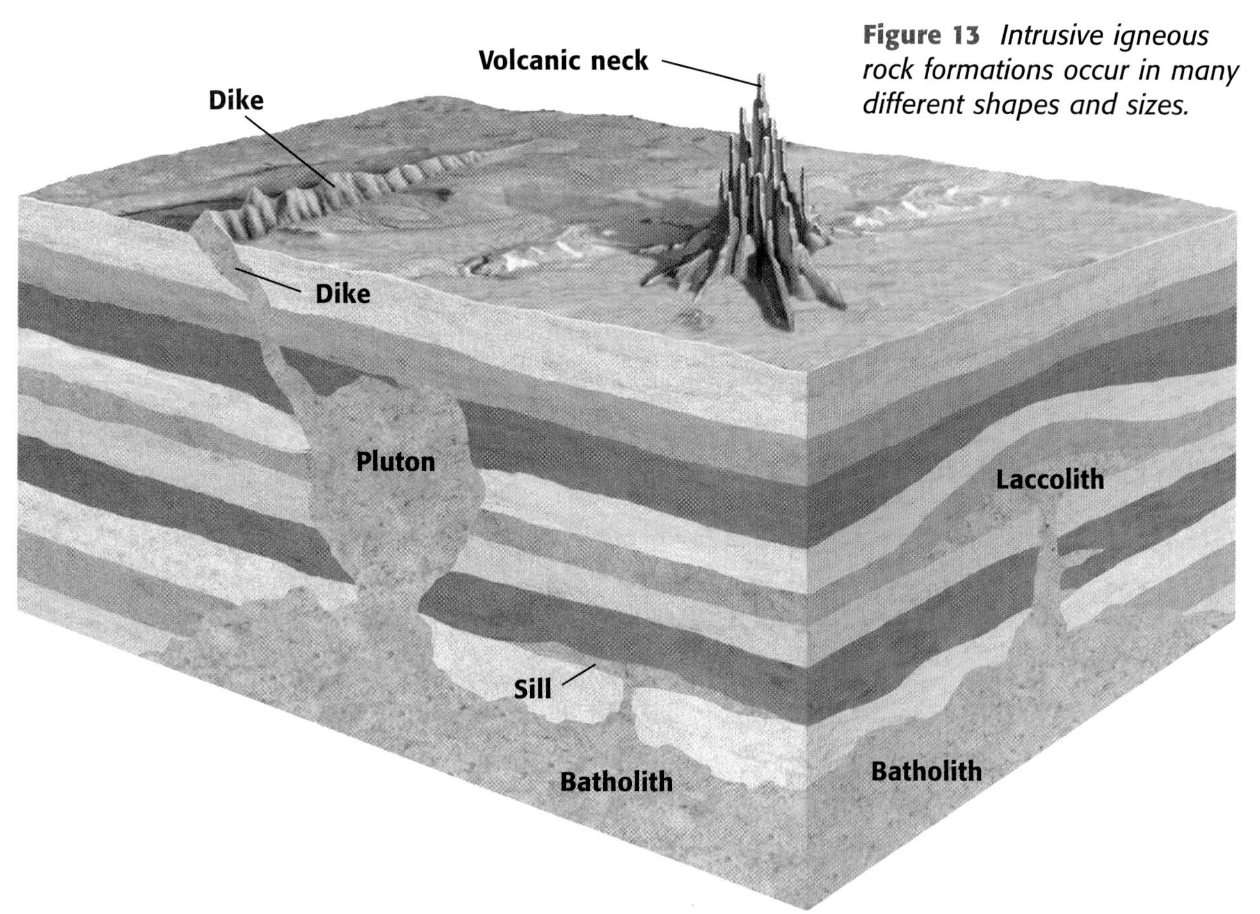

Figure 13 *Intrusive igneous rock formations occur in many different shapes and sizes.*

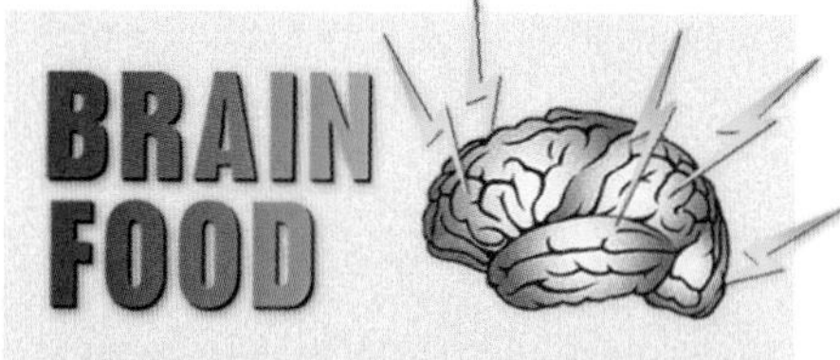

Violent volcanic eruptions sometimes produce a porous rock called pumice. Pumice is full of small holes once filled with trapped gases. Depending on how much space is taken up by these holes, some types of pumice can even float!

Extrusive Igneous Rock Igneous rock that forms on the Earth's surface is called **extrusive.** Most volcanic rock is extrusive. Extrusive rock cools quickly on the surface and contains either very small crystals or none at all.

When lava erupts from a volcano, a formation called a *lava flow* is made. You can see an active lava flow in **Figure 14.** But lava does not always come from volcanoes. Sometimes lava erupts from long cracks in the Earth's surface called *fissures.* When a large amount of lava flows out of a fissure, it can cover a vast area, forming a plain called a *lava plateau.* Pre-existing landforms are often buried by extrusive igneous rock formations.

Figure 14 *Below is an active lava flow. When exposed to surface conditions, lava quickly cools and solidifies, forming a fine-grained igneous rock.*

internet**connect**

SCILINKS NSTA

TOPIC: Igneous Rock
GO TO: www.scilinks.org
***sci*LINKS NUMBER:** HSTE093

SECTION REVIEW

1. What two properties are used to classify igneous rock?
2. How does the cooling rate of lava or magma affect the texture of an igneous rock?
3. **Interpreting Illustrations** Use the diagram in Figure 13 to compare a sill with a dike. What makes them different from each other?

READING WARM-UP

Sedimentary Rock

Terms to Learn

strata
stratification

What You'll Do

- Describe how the three types of sedimentary rock form.
- Explain how sedimentary rocks record Earth's history.

Wind, water, ice, sunlight, and gravity all cause rock to *weather* into fragments. **Figure 15** shows how some sedimentary rocks form. Through the process of erosion, rock fragments, called sediment, are transported from one place to another. Eventually the sediment is deposited in layers. Sedimentary rock then forms as sediments become compacted and cemented together.

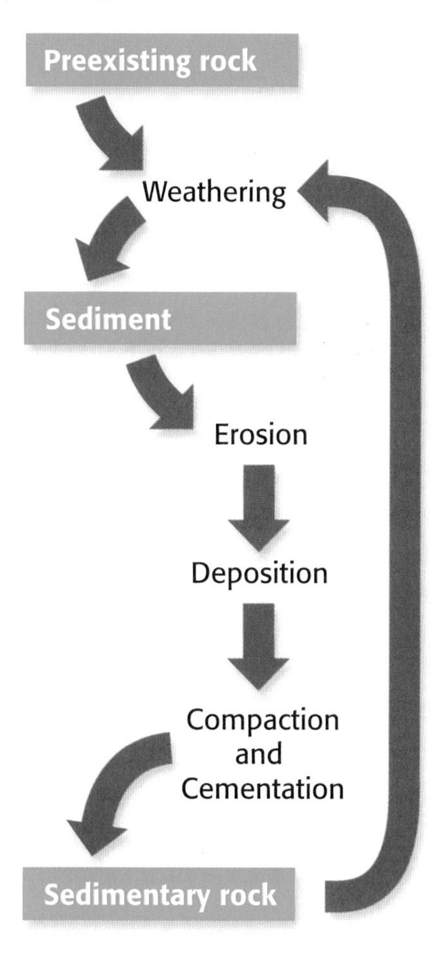

Figure 15 A Sedimentary Rock Cycle

Origins of Sedimentary Rock

As new layers of sediment are deposited, the layers eventually become compressed, or compacted. Dissolved minerals separate out of the water to form a natural glue that binds the sediments together into sedimentary rock. Sedimentary rock forms at or near the Earth's surface, without the heat and pressure involved in the formation of igneous and metamorphic rocks. The physical features of sedimentary rock tell part of its history. The most noticeable feature of sedimentary rock is its layers, or **strata.** Road cuts and construction zones are good places to observe sedimentary rock formations, and as you can see in **Figure 16,** canyons carved by rivers provide some spectacular views.

Figure 16 *Millions of years of erosion by the Colorado River have revealed the rock strata in the walls of the Grand Canyon.*

Composition of Sedimentary Rock

Sedimentary rock is also classified by the way it forms. There are three main categories of sedimentary rock—clastic, chemical, and organic. *Clastic* sedimentary rock forms when rock or mineral fragments, called clasts, stick together. *Chemical* sedimentary rock forms when minerals crystallize out of a solution, such as sea water, to become rock. *Organic* sedimentary rock forms from the remains of organisms.

Clastic Sedimentary Rock Clastic sedimentary rock is made of fragments of other rocks and minerals. As you can see in **Figure 17,** the size and shape of the rock fragments that make up clastic sedimentary rock influence their names.

Figure 17 *Clastic sedimentary rock is classified by the sizes of fragments it is made of.*

Chemical Sedimentary Rock Chemical sedimentary rock forms from *solutions* of minerals and water. As rainwater slowly makes its way to the ocean, it dissolves some of the rock material it passes through. Some of this dissolved material eventually forms the minerals that make up chemical sedimentary rock. One type of chemical sedimentary rock, chemical limestone, is made of calcium carbonate ($CaCO_3$), or the mineral calcite. It forms when calcium and carbonate become so concentrated in the sea water that calcite crystallizes out of the sea water solution, as shown in **Figure 18.**

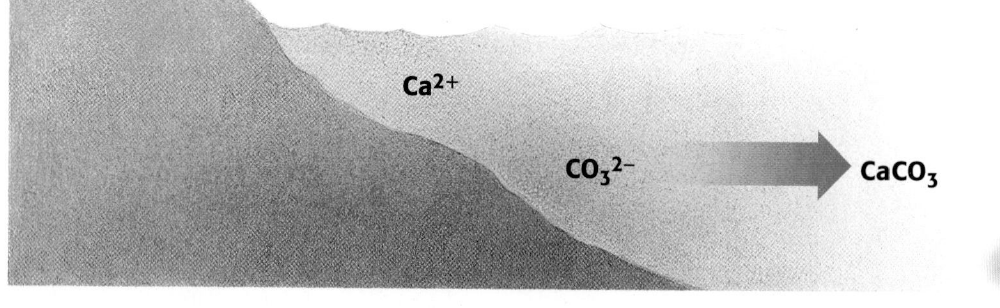

Figure 18 *Both salt water and fresh water contain dissolved calcium and carbonate. Chemical limestone forms on the ocean floor.*

Organic Sedimentary Rock

Most limestone forms from the remains of animals that once lived in the ocean. This organic material consists of shells or skeletons, which are made of calcium carbonate that the animals get from sea water.

For example, some limestone is made of the skeletons of tiny organisms called coral. Coral are very small, but they live in huge colonies, as shown in **Figure 19.** Over time, the remains of these sea animals accumulate on the ocean floor. These animal remains eventually become cemented together to form *fossiliferous* (FAHS uhl IF uhr uhs) *limestone.*

Figure 19 *Sea animals called coral create huge deposits of limestone. As they die, their skeletons accumulate on the ocean floor.*

Fossils are the remains or traces of plants and animals that have been preserved in sedimentary rock. Fossils have given us enormous amounts of information about ancient life-forms and how they lived. Most fossils come from animals that lived in the oceans. Another type of organic limestone, shown in **Figure 20,** forms from organisms that leave their shells in the mud on the ocean floor.

Figure 20 *Shellfish, such as clams (above right), get the calcium for their shells from sea water. When these organisms die, their shells collect on the ocean floor, eventually becoming rock (below). In time, huge rock formations result (right).*

Sedimentary Rock Structures

Many sedimentary rock features can tell you about the way the rock formed. The most characteristic feature of sedimentary rock is **stratification,** or layering. Strata differ from one another depending on the kind, size, and color of their sediment. The rate of deposition can also affect the thickness of the layers. Sedimentary rocks sometimes record the motion of wind and water waves on lakes, seas, rivers, and sand dunes. Some of these features are shown in **Figures 21** and **22.**

Figure 21 *Wind caused these slanted deposits, called* cross-beds, *but water can also cause them.*

Figure 22 *These* ripple marks *were made by flowing water and were preserved when the sediments became sedimentary rock. Ripple marks can also form from the action of wind.*

internet**connect**

TOPIC: Sedimentary Rock
GO TO: www.scilinks.org
***sci*LINKS NUMBER:** HSTE095

SECTION REVIEW

1. Describe the process by which clastic sedimentary rock forms.
2. List three sedimentary rock structures, and explain how they record geologic processes.
3. **Analyzing Relationships** Both clastic and chemical sedimentary rocks are classified according to texture and composition. Which property is more important for each sedimentary rock type? Explain.

SECTION 4

READING WARM-UP

Terms to Learn

foliated
nonfoliated

What You'll Do

- Describe two ways a rock can undergo metamorphism.
- Explain how the mineral composition of rocks changes as they undergo metamorphism.
- Describe the difference between foliated and nonfoliated metamorphic rock.

Metamorphic Rock

The word *metamorphic* comes from *meta,* meaning "changed," and *morphos,* meaning "shape." Remember, metamorphic rocks are those in which the structure, texture, or composition of the rock has changed. Rock can undergo metamorphism by heat or pressure acting alone or by a combination of the two. All three types of rock—igneous, sedimentary, and even metamorphic—can change into metamorphic rock.

Origins of Metamorphic Rock

The texture or mineral composition of a rock can change when its surroundings change. If the temperature or pressure of the new environment is different from the one the rock formed in, the rock will undergo metamorphism.

Most metamorphic change is caused by increased pressure that takes place at depths greater than 2 km. At depths greater than 16 km, the pressure can be more than 4,000 times the pressure of the atmosphere! Look at **Figure 23.** This rock, called garnet schist, formed at a depth of about 30 km. At this depth, some of the crystals the rock is made of change as a result of the extreme pressure. Other types of schist form at much shallower depths.

The temperature at which metamorphism occurs ranges from 50°C to 1,000°C. At temperatures higher than 1,000°C, most rocks will melt. Metamorphism does not melt rock—when rock melts, it becomes magma and then igneous rock. In **Figure 24** you can see that this rock was deformed by intense pressure.

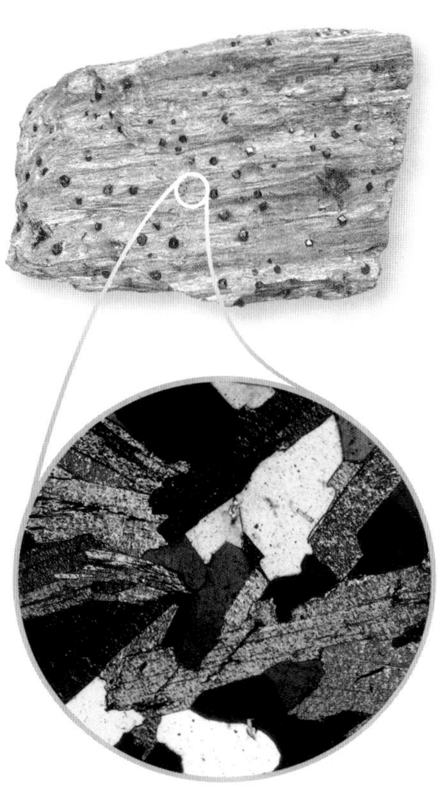

Figure 23 *At top is a metamorphic rock called garnet schist. At bottom is a microscopic view of a thin slice of a garnet schist.*

Figure 24 *In this outcrop, you can see an example of how sedimentary rock was deformed as it underwent metamorphism.*

QuickLab

Stretching Out

1. Draw your version of a granite rock on a **piece of paper** with a **black-ink pen.** Be sure to include the outline of the rock, and fill it in with different crystal shapes.
2. Mash some **plastic play putty** over the "granite," and slowly peel it off.
3. After making sure that the outline of your "granite" has been transferred to the putty, push and pull on the putty. What happened to the "crystals"? What happened to the "granite"?

TRY at HOME

Contact Metamorphism One way rock can undergo metamorphism is by coming into contact with magma. When magma moves through the crust, it heats the surrounding rock and "cooks" it. As a result, the magma changes some of the minerals in the surrounding rock into other minerals. The greatest change takes place where magma comes into direct contact with the surrounding rock. The effect of heat gradually lessens with distance from the magma. As you can see in **Figure 25,** *contact metamorphism* only happens next to igneous intrusions.

Regional Metamorphism When enormous pressure builds up in rock that is deeply buried under other rock formations, or when large pieces of the Earth's crust collide with each other, *regional metamorphism* occurs. The pressure and increased temperature that exist under these conditions cause rock to become deformed and chemically changed. This kind of metamorphic rock is underneath most continental rock formations.

Self-Check

How could a rock undergo both contact and regional metamorphism? *(See page 216 to check your answer.)*

Figure 25 *Metamorphism occurs over small areas, such as next to bodies of magma, and large areas, such as mountain ranges.*

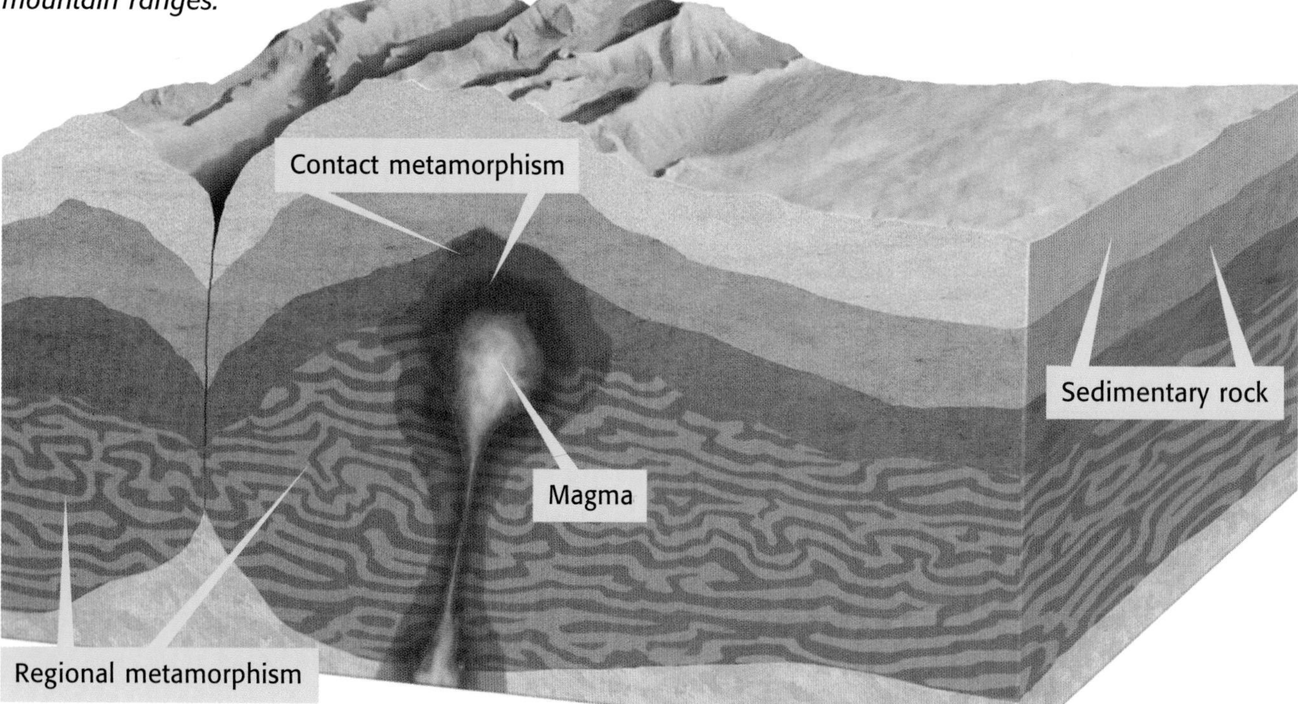

Composition of Metamorphic Rock

When conditions within the Earth's crust change because of collisions between continents or the intrusion of magma, the temperature and pressure of the existing rock change. Minerals that were present in the rock when it formed may no longer be stable in the new environment. The original minerals change into minerals that are more stable in the new temperature and pressure conditions. Look at **Figure 26** to see an example of how this happens.

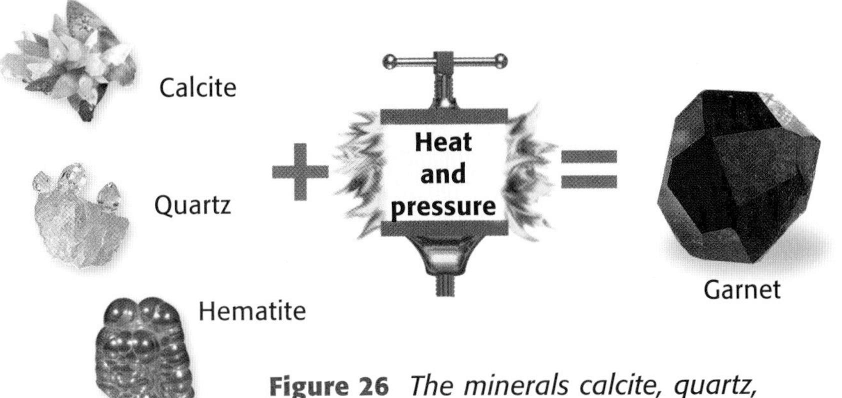

Figure 26 *The minerals calcite, quartz, and hematite combine and recrystallize to form the metamorphic mineral garnet.*

Activity

Did you know that you have a birthstone? Birthstones are gemstones, or mineral crystals. For each month of the year, there are one or two different birthstones. Find out which birthstone or birthstones you have by doing research in your school library or on the Internet. The names of birthstones are not usually the same as their actual mineral names. In what kind of rock would you likely find your birthstone? Why?

Many of these new minerals occur only in metamorphic rock. As shown in **Figure 27,** some metamorphic minerals form only within a specific range of temperature and pressure conditions. When scientists observe these metamorphic minerals in a rock, they can estimate the temperature and depth (pressure) at which recently exposed rock underwent metamorphism.

Figure 27 *Scientists can understand a metamorphic rock's history by observing the minerals it contains. For example, metamorphic rock containing garnet formed at a greater depth than one that contains only chlorite.*

Textures of Metamorphic Rock

As you know, texture helps to classify igneous and sedimentary rock. The same is true of metamorphic rock. All metamorphic rock has one of two textures—*foliated* or *nonfoliated.* **Foliated** metamorphic rock consists of minerals that are aligned and look almost like pages in a book. **Nonfoliated** metamorphic rock does not appear to have any regular pattern. Let's take a closer look at each of these types of metamorphic rock to find out how they form.

Foliated Metamorphic Rock Foliated metamorphic rock contains mineral grains that are aligned by pressure. Strongly foliated rocks usually contain flat minerals, like biotite mica. Look at **Figure 28.** Shale consists of layers of clay minerals. When subjected to slight heat and pressure, the clay minerals change into mica minerals and the shale becomes a fine-grained, foliated metamorphic rock called slate.

Metamorphic rocks can become other metamorphic rocks if the environment changes again. With additional heat and pressure, slate can change into phyllite, another metamorphic rock. When phyllite is exposed to additional heat and pressure, it can change into a metamorphic rock called schist.

As the degree of metamorphism increases, the arrangement of minerals in the rock changes. With additional heat and pressure, coarse-grained minerals separate into bands in a metamorphic rock called *gneiss* (pronounced "nice").

Sedimentary shale

Slate

Phyllite

Schist

Gneiss

Figure 28 *The effects of metamorphism depend on the heat and pressure applied to the rock. Here you can see what happens to shale when it is exposed to more and more heat and pressure.*

Wouldn't it be "gneiss" to make your own foliated rock? Turn to page 181 in your LabBook to find out how.

Nonfoliated Metamorphic Rock Nonfoliated metamorphic rocks are shown in **Figure 29.** Do you notice anything missing? The lack of aligned mineral grains makes them nonfoliated. They are rocks commonly made of only one, or just a few, minerals.

Sandstone is a sedimentary rock made of distinct quartz sand grains. But when sandstone is subjected to the heat and pressure of metamorphism, the spaces between the sand grains disappear as they recrystallize, forming quartzite. Quartzite has a shiny, glittery appearance. It is still made of quartz, but the mineral grains are larger. When limestone undergoes metamorphism, the same process happens to the mineral calcite, and the limestone becomes marble. Marble has larger calcite crystals than limestone. You have probably seen marble in buildings and statues.

The term *metamorphosis* means "change in form." When certain animals undergo a dramatic change in the shape of their body, they are said to have undergone a metamorphosis. As part of their natural life cycle, moths and butterflies go through four stages of life. After they hatch from an egg, they are in the larval stage in the form of a caterpillar. In the next stage they build a cocoon or become a chrysalis. This is called the pupal stage. They finally emerge into the adult stage of their life, complete with wings, antennae, and legs!

Marble

Quartzite

Figure 29 *Marble and quartzite are nonfoliated metamorphic rocks. As you can see in the microscopic views, none of the mineral crystals are aligned.*

SECTION REVIEW

1. What environmental factors cause rock to undergo metamorphism?
2. What is the difference between foliated and nonfoliated metamorphic rock?
3. **Making Inferences** If you had two metamorphic rocks, one with garnet crystals and the other with chlorite crystals, which one would have formed at a deeper level in the Earth's crust? Explain.

TOPIC: Metamorphic Rock
GO TO: www.scilinks.org
***sci*LINKS NUMBER:** HSTE098

Making Models Lab

Round and Round in Circles

The rocks that make up the Earth are constantly being recycled. One form of rock is often broken down and changed into another form of rock. Do this activity to learn what happens to rocks as they change from one rock type to another.

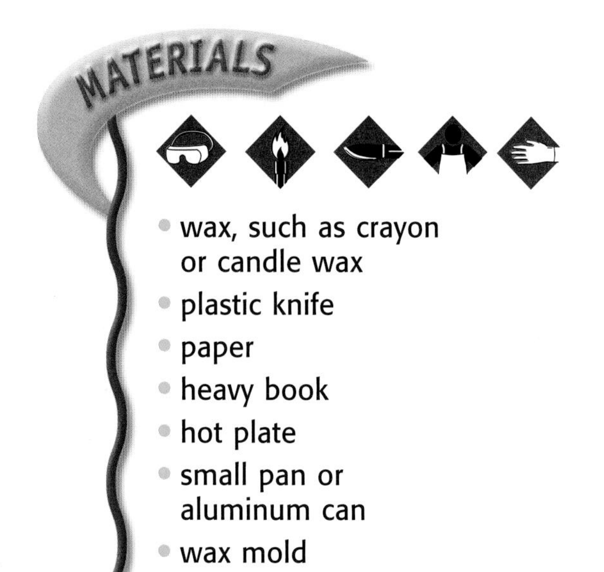

- wax, such as crayon or candle wax
- plastic knife
- paper
- heavy book
- hot plate
- small pan or aluminum can
- wax mold

Procedure

1. Take several pieces of wax of different colors. Carefully scrape off some wax of each color with the edge of a plastic knife. These shavings will represent tiny grains of rock or sand.

2. When you have made a large pile of wax shavings, cover it with a piece of paper and a heavy book. Gently press down on the book until the wax shavings stick together. This mixture of mineral grains will represent sedimentary rock. Write a description of the "rock" in your ScienceLog. How is the rock different from the tiny shavings you started with?

3. Now take your wax "sedimentary rock" and warm it in your hands for a while. Place the paper and the book on top of the warm wax. Press down on the wax a little harder than you did in the first step. Fold the warmed wax in half, and press down some more. This second type of "rock" represents metamorphic rock. Describe this new "rock" in your ScienceLog.

4. Place the wax in the pan. Turn on the hot plate, and place the pan on the hot plate. Observe the wax as it melts. In this state, what does the wax represent?

5. Turn off the hot plate. Carefully pour the melted wax into a mold. Observe the wax as it cools and hardens. Carefully touch the wax with the eraser end of your pencil. Record your observations. The newly cooled wax represents igneous rock. Describe how this "rock" is different from the first two.

6. Finally, take the cooled wax and scrape off bits of the "rock" to form grains of rock or sand. If you have time, repeat steps 2–5.

Analysis

7. Review your descriptions of each type of "rock." Which "rock" do you think resembles the rock that forms from erupting volcanoes?

8. Which "rock" is formed from pieces of broken-down rock? How do these rock fragments harden into rock?

9. In step 6 you were asked to go back to step 2. Explain how this activity can be described as a rock cycle.

10. Does this model of the rock cycle have any limitations? Explain your answer.

Chapter Highlights

SECTION 1

Vocabulary

rock *(p. 26)*
rock cycle *(p. 28)*
magma *(p. 29)*
sedimentary rock *(p. 30)*
metamorphic rock *(p. 30)*
igneous rock *(p. 30)*
composition *(p. 31)*
texture *(p. 32)*

Section Notes

- Rocks have been used by humans for thousands of years, and they are just as valuable today.
- Rocks are classified into three main types—igneous, sedimentary, and metamorphic—depending on how they formed.
- The rock cycle describes the process by which a rock can change from one rock type to another.
- Scientists further classify rocks according to two criteria—composition and texture.
- Molten igneous material creates rock formations both below and above ground.

SECTION 2

Vocabulary

intrusive *(p. 35)*
extrusive *(p. 36)*

Section Notes

- The texture of igneous rock is determined by the rate at which it cools. The slower magma cools, the larger the crystals are.
- Felsic igneous rock is light-colored and lightweight, while mafic igneous rock is dark-colored and heavy.
- Igneous material that solidifies at the Earth's surface is called extrusive, while igneous material that solidifies within the crust is called intrusive.

Lab

Crystal Growth *(p. 178)*

Skills Check

Math Concepts

MINERAL COMPOSITION Rocks are classified not only by the minerals they contain but also by the amounts of those minerals. Suppose a particular kind of granite is made of feldspar, biotite mica, and quartz. If you know that feldspar makes up 55 percent of the rock and biotite mica makes up 15 percent of the rock, the remaining 30 percent must be made of quartz.

55% feldspar		100% of granite
+ 15% biotite mica	or	– 55% feldspar
+ 30% quartz		– 15% biotite mica
= 100% of granite		= 30% quartz

Visual Understanding

PIE CHARTS The pie charts on page 31 help you visualize the relative amounts of minerals in different types of rock. The circle represents the whole rock, or 100 percent. Each part, or "slice," of the circle represents a fraction of the rock.

10% Biotite mica
35% Quartz
55% Feldspar

SECTION 3

Vocabulary

strata *(p. 37)*
stratification *(p. 40)*

Section Notes

- Clastic sedimentary rock is made of rock and mineral fragments that are compacted and cemented together. Chemical sedimentary rock forms when minerals crystallize out of a solution such as sea water. Organic sedimentary rock forms from the remains of organisms.
- Sedimentary rocks record the history of their formation in their features. Some common features are strata, ripple marks, and fossils.

Lab

Let's Get Sedimental *(p. 182)*

SECTION 4

Vocabulary

foliated *(p. 44)*
nonfoliated *(p. 44)*

Section Notes

- One kind of metamorphism is the result of magma heating small areas of surrounding rock, changing its texture and composition.
- Most metamorphism is the product of heat and pressure acting on large regions of the Earth's crust.
- The mineral composition of a rock changes when the minerals it is made of recrystallize to form new minerals. These new minerals are more stable under increased temperature and pressure.
- Metamorphic rock that contains aligned mineral grains is called foliated, and metamorphic rock that does not contain aligned mineral grains is called nonfoliated.

Lab

Metamorphic Mash *(p. 181)*

internet**connect**

GO TO: go.hrw.com

Visit the **HRW** Web site for a variety of learning tools related to this chapter. Just type in the keyword:

KEYWORD: HSTRCK

GO TO: www.scilinks.org

Visit the **National Science Teachers Association** on-line Web site for Internet resources related to this chapter. Just type in the ***sci*LINKS** number for more information about the topic:

TOPIC: Composition of Rock — ***sci*LINKS NUMBER:** HSTE090
TOPIC: Igneous Rock — ***sci*LINKS NUMBER:** HSTE093
TOPIC: Sedimentary Rock — ***sci*LINKS NUMBER:** HSTE095
TOPIC: Metamorphic Rock — ***sci*LINKS NUMBER:** HSTE098
TOPIC: Rock Formations — ***sci*LINKS NUMBER:** HSTE100

Chapter Review

USING VOCABULARY

To complete the following sentences, choose the correct term from each pair of terms listed below:

1. ___?___ igneous rock is more likely to have coarse-grained texture than ___?___ igneous rock. (*Extrusive/intrusive* or *Intrusive/extrusive*)

2. ___?___ metamorphic rock texture consists of parallel alignment of mineral grains. (*Foliated* or *Nonfoliated*)

3. ___?___ sedimentary rock forms when grains of sand become cemented together. (*Clastic* or *Chemical*)

4. ___?___ cools quickly on the Earth's surface. (*Lava* or *Magma*)

5. Strata are found in ___?___ rock. (*igneous* or *sedimentary*)

UNDERSTANDING CONCEPTS

Multiple Choice

6. A type of rock that forms deep within the Earth when magma solidifies is called
   a. sedimentary.
   b. metamorphic.
   c. organic.
   d. igneous.

7. A type of rock that forms under high temperature and pressure but is not exposed to enough heat to melt the rock is
   a. sedimentary.
   b. metamorphic.
   c. organic.
   d. igneous.

8. After they are deposited, sediments, such as sand, are turned into sedimentary rock when they are compacted and
   a. cemented.
   b. metamorphosed.
   c. melted.
   d. weathered.

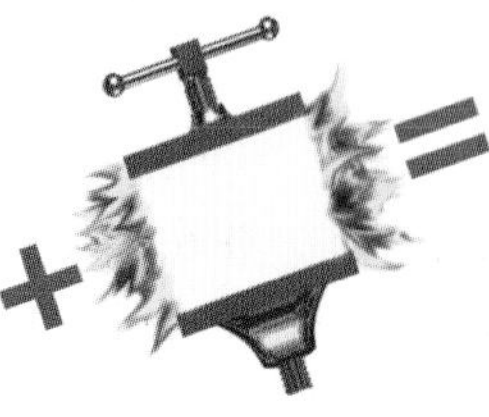

9. An igneous rock with a coarse-grained texture forms when
   a. magma cools very slowly.
   b. magma cools very quickly.
   c. magma cools quickly, then slowly.
   d. magma cools slowly, then quickly.

10. The layering that occurs in sedimentary rock is called
    a. foliation.
    b. ripple marks.
    c. stratification.
    d. compaction.

11. An example of a clastic sedimentary rock is
    a. obsidian.
    b. sandstone.
    c. gneiss.
    d. marble.

12. A common sedimentary rock structure is
    a. a sill.
    b. a pluton.
    c. cross-bedding.
    d. a lava flow.

13. An example of mafic igneous rock is
    a. granite.
    b. basalt.
    c. quartzite.
    d. pumice.

14. Chemical sedimentary rock forms when
    a. magma cools and solidifies.
    b. minerals are twisted into a new arrangement.
    c. minerals crystallize from a solution.
    d. sand grains are cemented together.

15. Which of the following is a foliated metamorphic rock?
    a. sandstone
    b. gneiss
    c. shale
    d. basalt

Short Answer

16. In no more than three sentences, explain the rock cycle.

17. How are sandstone and siltstone different from one another? How are they the same?

18. In one or two sentences, explain how the cooling rate of magma affects the texture of the igneous rock that forms.

Concept Mapping

19. Use the following terms to create a concept map: rocks, clastic, metamorphic, nonfoliated, igneous, intrusive, chemical, foliated, organic, extrusive, sedimentary.

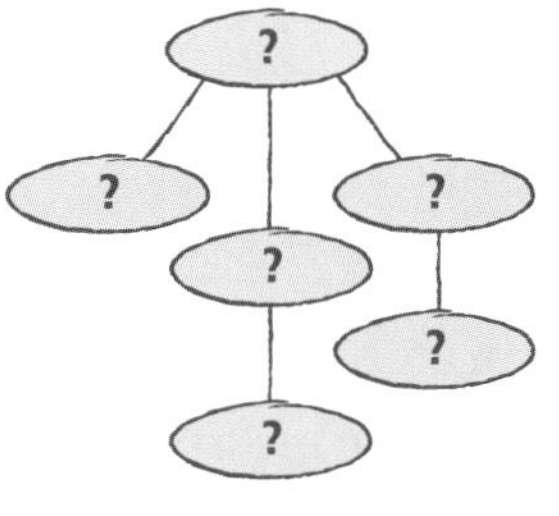

CRITICAL THINKING AND PROBLEM SOLVING

Write one or two sentences to answer the following questions:

20. The sedimentary rock coquina is made up of pieces of seashells. Which of the three kinds of sedimentary rock could it be? Explain.

21. If you were looking for fossils in the rocks around your home and the rock type that was closest to your home was metamorphic, would you find many fossils? Why or why not?

22. Suppose you are writing a book about another planet. In your book, you mention that the planet has no atmosphere or weather. Which type of rock will you not find on the planet? Explain.

23. Imagine that you want to quarry or mine granite. You have all of the equipment, but you need a place to quarry. You have two pieces of land to choose from. One piece is described as having a granite batholith under it, and the other has a granite sill. If both plutonic bodies were at the same depth, which one would be a better buy for you? Explain your answer.

MATH IN SCIENCE

24. If a 60 kg granite boulder were broken down into sand grains and if quartz made up 35 percent of the boulder's mass, how many kilograms of the resulting sand would be quartz grains?

INTERPRETING GRAPHICS

The red curve on the graph below shows how the melting point of a particular rock changes with increasing temperature and pressure. Use the graph to answer the questions below.

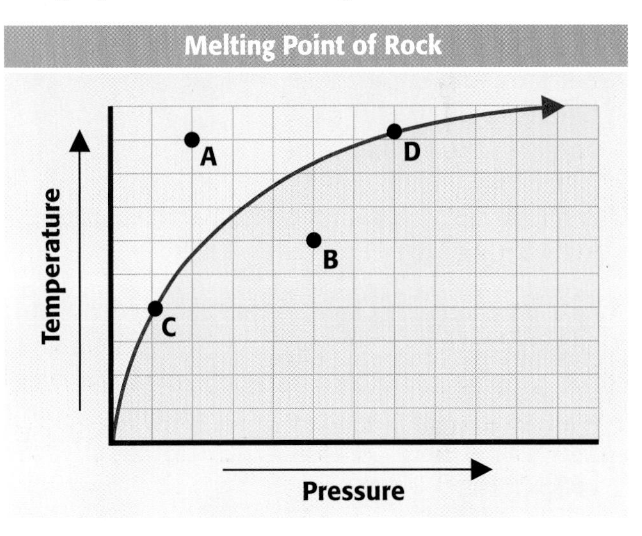

25. What type of material, liquid or solid, would you find at point **A**? Why?

26. What would you find at point **B**?

27. Points **C** and **D** represent different temperature and pressure conditions for a single, solid rock. Why does this rock have a higher melting temperature at point **D** than it does at point **C**?

Take a minute to review your answers to the Pre-Reading Questions found at the bottom of page 24. Have your answers changed? If necessary, revise your answers based on what you have learned since you began this chapter.

Science, Technology, and Society

Rock City

Today when we dig into a mountainside to build a highway or make room for a building, we use heavy machinery and explosives. Can you imagine doing the same job with just a hammer and chisel? Well, between about 300 B.C. and A.D. 200, an Arab tribe called the Nabataeans (nab uh TEE uhns) did just that. In fact, they carved a whole city—homes, storage areas, monuments, administrative offices, and temples—right into the mountainsides!

▲ *Petra's most famous building, the Treasury, was shown in the movie* Indiana Jones and the Last Crusade.

Rose-Red City

This amazing city in southern Jordan is Petra (named by the Roman emperor Hadrian Petra during a visit in A.D. 131). A poet once described Petra as "the rose-red city" because all the buildings and monuments were carved from the pink sandstone mountains surrounding Petra.

Using this reddish stone, the Nabataeans lined the main street in the center of the city with tall stone columns. The street ends at what was once the foot of a mountain but is now known as the Great Temple—a two-story stone religious complex larger than a football field!

The High Place of Sacrifice, another site near the center of the city, was a mountaintop. The Nabataeans leveled the top and created a place of worship more than 1,000 m above the valley floor. Today visitors climb stairs to the top. Along the way, they pass dozens of tombs carved into the pink rock walls.

Tombs and More Tombs

There are more than 800 other tombs dug into the mountainsides in and around Petra. One of them, the Treasury (created for a Nabataean ruler), stands more than 40 m high! It is a magnificent building with an elaborate facade. Behind the massive stone front, the Nabataeans carved one large room and two smaller rooms deeper into the mountain.

Petra Declines

The Nabataeans once ruled an area extending from Petra to Damascus. They grew wealthy and powerful by controlling important trade routes near Petra. But their wealth attracted the Roman Empire, and in A.D. 106, Petra became a Roman province. Though the city prospered under Roman rule for almost another century, a gradual decline in Nabataean power began. The trade routes by land that the Nabataeans controlled for hundreds of years were abandoned in favor of a route by the Red Sea. People moved and the city faded. By the seventh century, nothing was left of Petra but empty stone structures.

Think About It!

► Petra is sometimes referred to as a city "from the rock as if by magic grown." Why might such a city seem "magic" to us today? What might have encouraged the Nabataeans to create this city? Share your thoughts with a classmate.

Glass Scalpels

Would you want your surgeon to use a scalpel that was thousands of years old? Probably not, unless it was a razor-sharp knife blade made of obsidian, a natural volcanic glass. Such blades and arrowheads were used for nearly 18,000 years by our ancestors. Recently, physicians have found a new use for these Stone Age tools. Obsidian blades, once used to hunt woolly mammoths, are now being used as scalpels in the operating room!

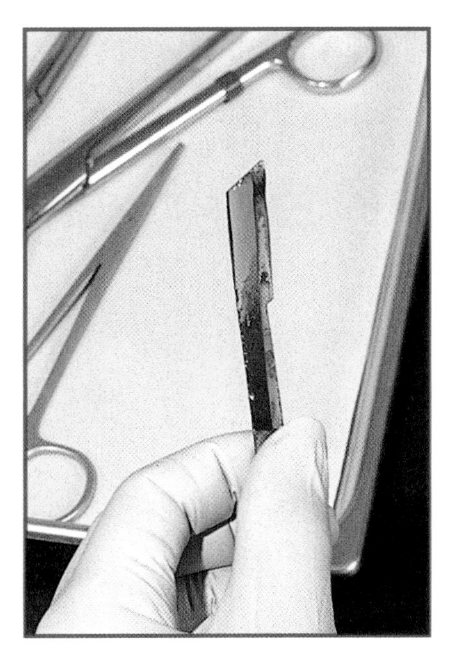

▲ *An obsidian scalpel can have an edge as fine as a single molecule.*

Obsidian or Stainless Steel?

Traditionally, physicians have used inexpensive stainless-steel scalpel blades for surgical procedures. Steel scalpels cost about $2 each, and surgeons use them just once and throw them away. Obsidian scalpels are more expensive—about $20 each—but they can be used many times before they lose their keen edge. And obsidian scalpel blades can be 100 times sharper than traditional scalpel blades!

During surgery, steel scalpels actually tear the skin apart. Obsidian scalpels divide the skin and cause much less damage. Some plastic surgeons use obsidian blades to make extremely fine incisions that leave almost no scarring. An obsidian-scalpel incision heals more quickly because the blade causes less damage to the skin and other tissues.

Many patients have allergic reactions to mineral components in steel blades. These patients often do not have an allergic reaction when obsidian scalpels are used. Given all of these advantages, it is not surprising that some physicians have made the change to obsidian scalpels.

A Long Tradition

Early Native Americans were among the first people to recognize that chipped obsidian has extremely sharp edges. Native Americans made obsidian arrowheads and knife blades by flaking away chips of rock by hand. Today obsidian scalpels are fashioned in much the same way by a *knapper,* a person who makes stone tools by hand. Knappers use the same basic technique that people have used for thousands of years to make obsidian blades and other stone tools.

Find Out for Yourself!

► Making obsidian blades and other stone tools requires a great deal of skill. Find out about the steps a knapper follows to create a stone tool. Find a piece of rock, and see if you can follow the steps to create a stone tool of your own. Be careful not to hit your fingers, and wear safety goggles.

CHAPTER 3

The Rock and Fossil Record

Sections

1. How can you determine if some rocks and fossils are older than others?
2. Are fossils always made up of parts of plants or animals?
3. How do scientists study the Earth's history?

TIME STANDS STILL

Sealed in darkness for 49 million years, this beetle still shimmers with the same metallic hues that once helped it hide among ancient plants. This rare fossil was found in Messel, Germany. In the same rock formation, scientists have found fossilized crocodiles, bats, birds, and frogs. A living stag beetle *(below)* has a similar form and color. Do you think that these two beetles would live in similar environments? What do you think Messel, Germany, was like 49 million years ago? In this chapter, you will learn how scientists answer questions like these.

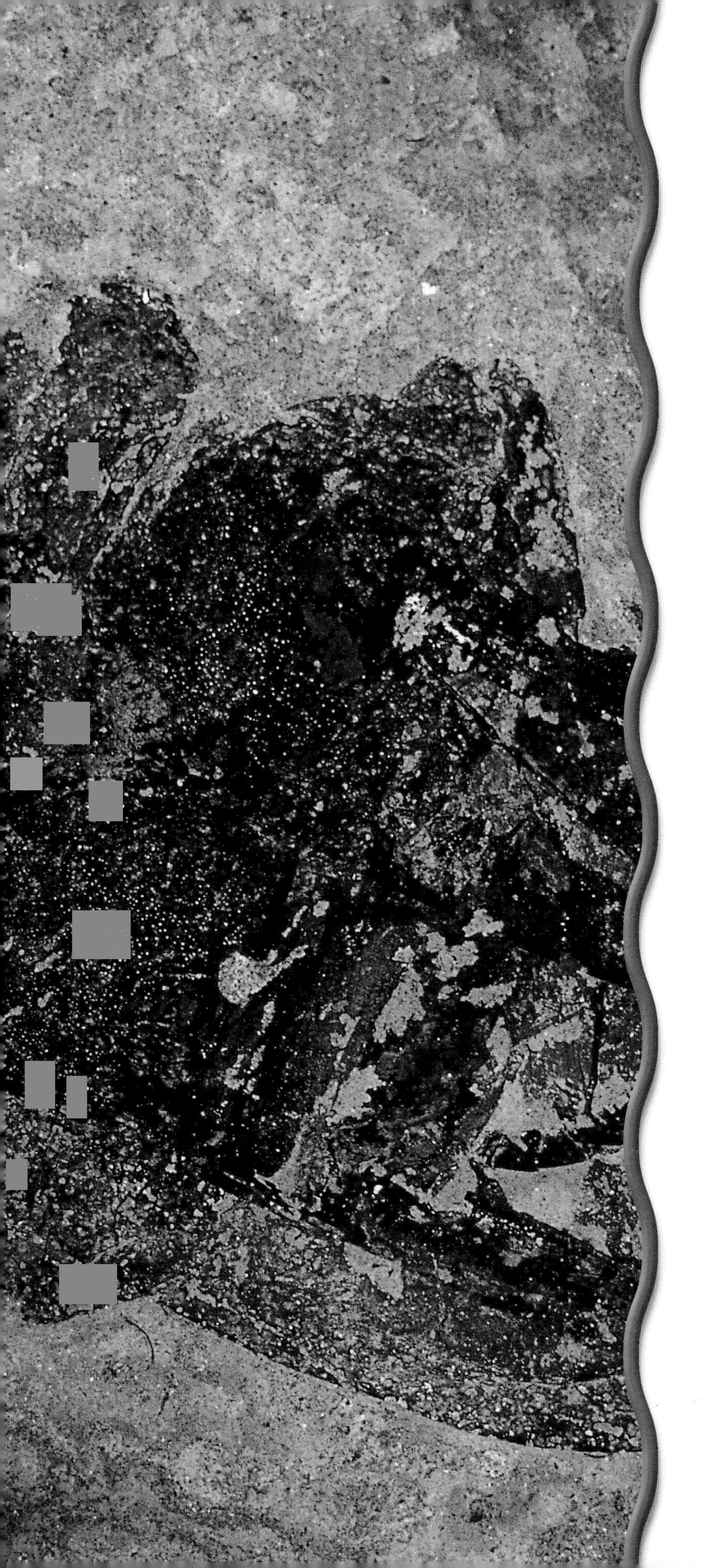

START-UP Activity

MAKING FOSSILS

How do scientists learn from fossils? In this activity, you will study "fossils" and identify the object that made each.

Procedure

1. You and three or four of your classmates will be given several pieces of **modeling clay** and a paper sack containing a few **small objects.**
2. Press each object firmly into a piece of clay. Try to leave a fossil imprint showing as much detail as possible.
3. After you have made an imprint of each object, exchange your model fossils with another group.
4. In your ScienceLog, describe the fossils you have received. List as many details as possible. What patterns and textures do you observe?
5. Work as a group to identify each fossil and check your results. Were you right?

Analysis

6. What kinds of details were important in identifying your fossils? What kinds of details were not preserved in the imprints? For example, can you tell the color of the objects?
7. Explain how Earth scientists follow similar methods when studying fossils.

SECTION 1

READING WARM-UP

Terms to Learn

uniformitarianism
catastrophism

What You'll Do

- Identify the role of uniformitarianism in Earth science.
- Contrast uniformitarianism with catastrophism.
- Describe how the role of catastrophism in Earth science has changed.

Earth's Story and Those Who First Listened

Humans have wondered about Earth's history for thousands of years. But the branch of Earth science called *geology,* which involves the study of Earth's history, got a late start. The main concept of modern geology was not outlined until the late eighteenth century. Within a few decades, this concept replaced a more traditional concept of Earth's history. Today, both concepts are an essential part of Earth science.

The Principle of Uniformitarianism

In 1795, a philosopher and scientist named James Hutton published *Theory of the Earth,* in which he wrote that Earth's landforms are constantly changing. As shown in **Figure 1,** Hutton assumed that these changes result from geologic processes—such as the breakdown of rock and the transport of sediment—that remain uniform, or do not change, over time. This assumption is now called uniformitarianism. **Uniformitarianism** is a principle that states that the same geologic processes shaping the Earth today have been at work throughout Earth's history. "The present is the key to the past" is a phrase that best summarizes uniformitarianism.

Figure 1 *Hutton observed gradual, uniform geologic processes at work. Judging by the slowness of the processes, he concluded that the Earth must be incredibly old.*

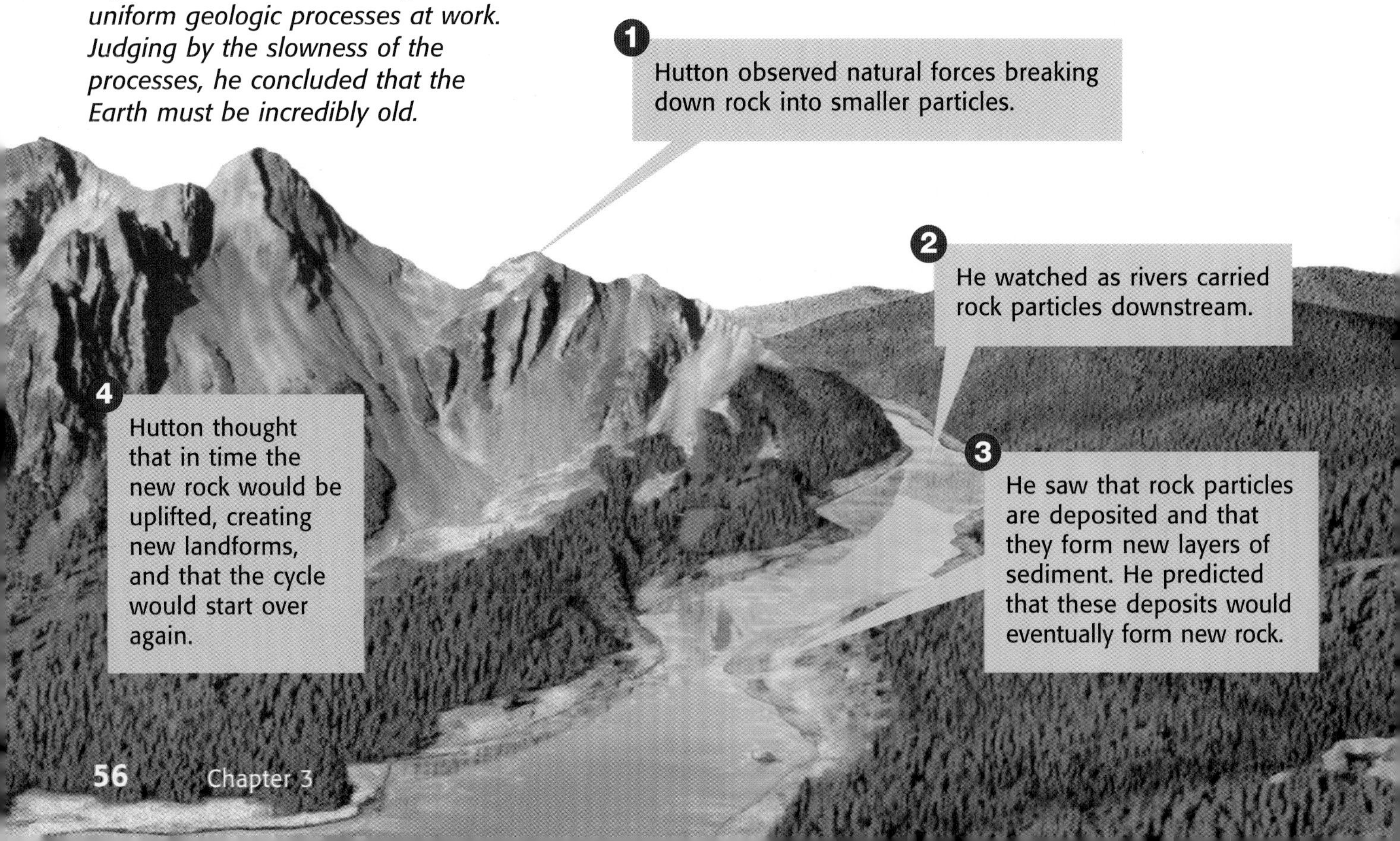

Making Assumptions

Examine the photographs at right. List the letters of the photos in the order you think the photos were taken. Now think of all the assumptions that you made to infer that order. Write down as many of these assumptions as you can. Compare notes with your classmates. Did you get the same sequence? Were your assumptions similar?

In science, assumptions must also be made. For example, you assume that the sun will rise each day. Briefly explain the importance of being able to count on certain things always being the same. How does this apply to uniformitarianism?

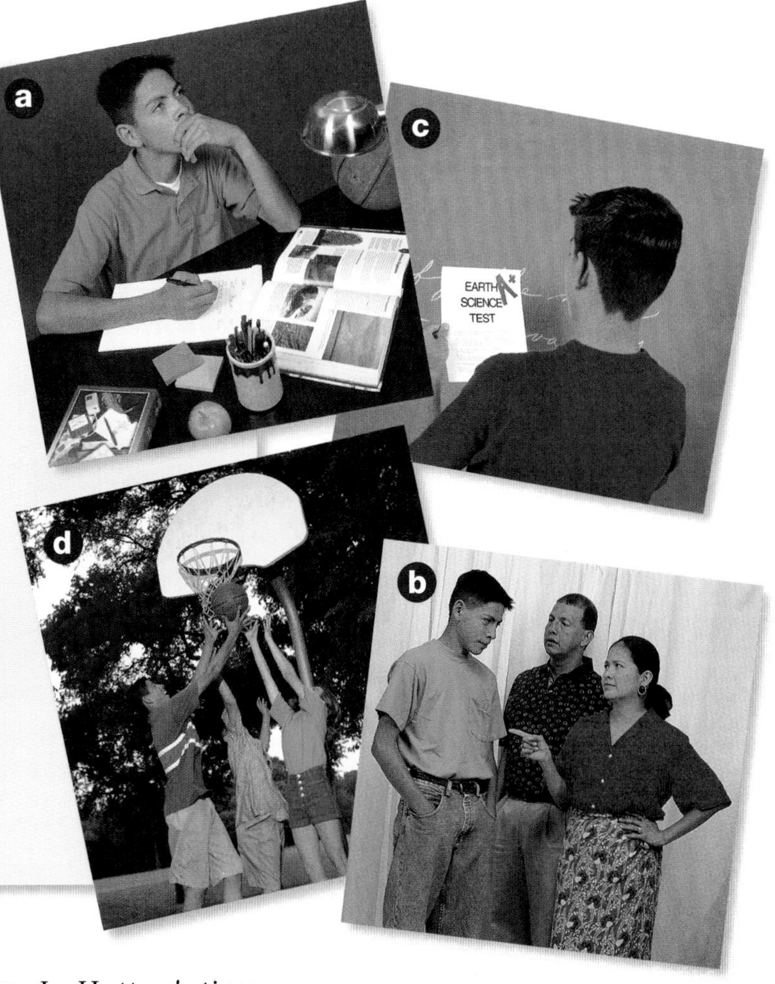

Uniformitarianism Versus Catastrophism In Hutton's time most people thought that the Earth had existed for only thousands of years. This was not nearly enough time for the gradual geologic processes that Hutton described to have shaped our planet. But uniformitarianism was not immediately accepted. Instead, most scientists believed in catastrophism. **Catastrophism** is a principle that states that all geologic change occurs suddenly. Supporters of catastrophism claimed that the formation of all Earth's features, such as its mountains, canyons, and seas, could be explained by rare, sudden events called *catastrophes*. These unpredictable catastrophes caused rapid geologic changes over large areas—sometimes even globally.

Uniformitarianism Wins! Despite Hutton's observations, catastrophism remained geology's guiding principle for decades. It took the work of Charles Lyell, another scientist, for people to seriously consider uniformitarianism.

From 1830 to 1833, Lyell published three volumes collectively titled *Principles of Geology,* in which he reintroduced uniformitarianism. Armed with Hutton's notes and new evidence of his own, Lyell successfully challenged the principle of catastrophism. Lyell saw no reason to doubt that major geologic change happened the same way in the past as it does in the present—gradually.

As a friend of Charles Lyell, Charles Darwin was greatly influenced by Lyell's uniformitarian ideas. Lyell's influence became clear when Darwin published *On the Origin of Species by Natural Selection* in 1859. Similar to uniformitarianism, Darwin's theory of evolution proposes that changes in species occur gradually over long periods of time.

Modern Geology—A Happy Medium

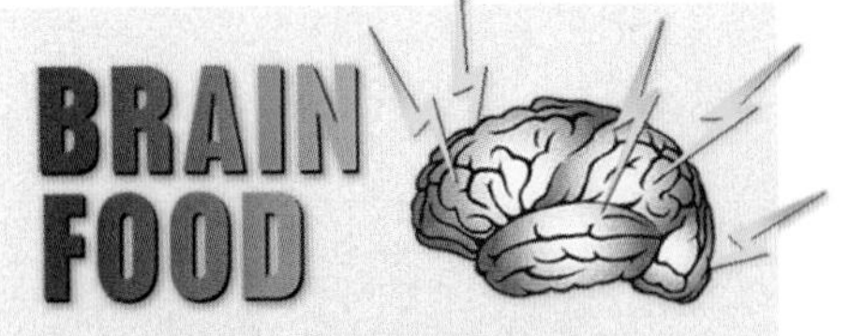

Did you know that the first dinosaur bones were not identified until 1841? Hutton and Lyell developed their ideas without knowledge of these giants of prehistory.

Today scientists realize that neither uniformitarianism nor catastrophism accounts for all of Earth's history. Although most geologic change is gradual and uniform, catastrophes do occur occasionally. For example, huge craters have been found where asteroids and comets are thought to have struck Earth in the past. Some of these strikes indeed may have been catastrophic. Some scientists think one such asteroid strike led to the extinction of the dinosaurs, as explained in **Figure 2.** The impact of an asteroid is thought to have spread debris into the atmosphere around the entire planet, blocking the sun's rays and causing major changes in the global climate.

Figure 2 *Today scientists think that sudden events are responsible for some changes in Earth's past. An asteroid hitting Earth, for example, may have led to the extinction of the dinosaurs 65 million years ago.*

internet**connect**

TOPIC: Earth's Story
GO TO: www.scilinks.org
***sci*LINKS NUMBER:** HSTE130

SECTION REVIEW

1. Why do Earth scientists need the principle of uniformitarianism in order to make predictions?
2. What is the difference between uniformitarianism and catastrophism?
3. **Summarizing Data** How has the role of catastrophism in Earth science changed?

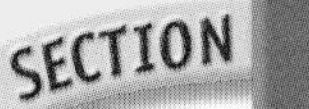

SECTION 2

READING WARM-UP

Terms to Learn

relative dating
superposition
geologic column
unconformity

What You'll Do

- Explain how relative dating is used in geology.
- Explain the principle of superposition.
- Demonstrate an understanding of the geologic column.
- Identify two events and two features that disrupt rock sequences.
- Explain how physical features are used to determine relative ages.

Relative Dating: Which Came First?

Imagine that you are a detective investigating a crime scene. What is the first thing you would do? You might begin by dusting the scene for fingerprints or by searching for witnesses. As a detective, your goal is to figure out the sequence of events that took place before you arrived at the scene.

Geologists have a similar goal when investigating the Earth. They try to determine the order of events that led to how the Earth looks today. But instead of fingerprints and witnesses, geologists rely on rocks and fossils. Determining whether an object or event is older or younger than other objects or events is called **relative dating.**

The Principle of Superposition

Suppose you have an older brother who takes a lot of photographs of your family but never puts them into an album. He just piles them in a box. Over the years, he keeps adding new pictures to the top of the stack. Think about the family history recorded in those pictures. Where are the oldest pictures—the ones taken when you were a baby? Where are the most recent pictures—those taken last week?

Rock layers, such as the ones shown in **Figure 3,** are like stacked pictures. The oldest layers are at the bottom. As you move from bottom to top, the layers get more recent, or younger. Scientists call this superposition. **Superposition** is a principle that states that younger rocks lie above older rocks in undisturbed sequences. "Younger over older" is a phrase you can use to remember this principle.

Figure 3 *Rock layers are like photos stacked over time—the younger ones lie above the older ones.*

Activity

1. Write the titles of 10 chapters of this book on 10 note cards (one title on each note card).
2. Shuffle the cards and exchange them with a partner. Try to put your partner's titles in the correct order without using your book.
3. Compare your order with the order in the book.
4. Your work would have been easier if you had been allowed to use your book. How does this relate to geologists using the geologic column to put rock layers in order?

Disturbing Forces Some rock-layer sequences, however, are disturbed by forces from within the Earth. These forces can push other rocks into a sequence, tilt or fold rock layers, and break sequences into movable parts. Sometimes these forces even put older layers above younger layers, which goes against superposition. The disruptions of rock sequences caused by these forces pose a great challenge to geologists trying to determine the relative ages of rocks. Fortunately, geologists can get help from a very valuable tool—the geologic column.

The Geologic Column

To make their job easier, geologists combine data from all the known undisturbed rock sequences around the world. From this information, geologists create the *geologic column.* The **geologic column** is an ideal sequence of rock layers that contains all the known fossils and rock formations on Earth arranged from oldest to youngest.

Geologists rely on the geologic column to interpret rock sequences. For example, when geologists are not sure about the age of a rock sequence they are studying, they gather information about the sequence and compare it to the geologic column. Geologists also use the geologic column to identify the layers in puzzling rock sequences, such as sequences that have been folded over.

Constructing the Geologic Column

Here you can see three rock sequences (**a, b,** and **c**) from three different locations. Some rock layers appear in more than one sequence. Geologists construct the geologic column by piecing together different rock sequences from all over the world.

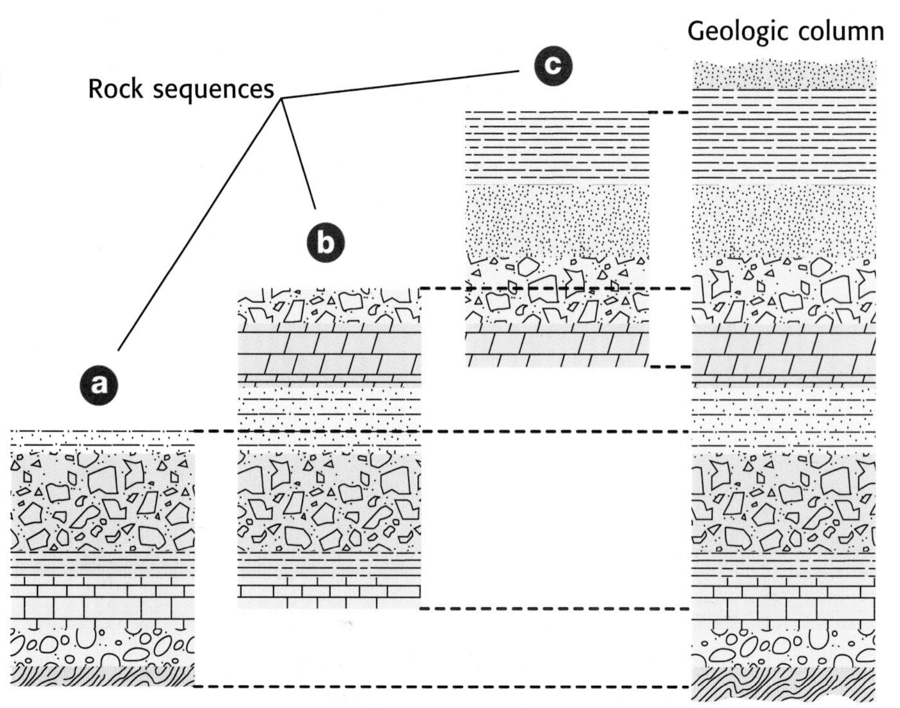

Disturbed Rock Layers

Geologists often find features that cut through existing rock layers. Geologists use the relationships between rock layers and the features that cut across them to assign relative ages to the features and the layers. They know that those features are younger than the rock layers because the rock layers had to be present before the features could cut across them.

Faults and intrusions are examples of features that cut across rock layers. A *fault* is a break in the Earth's crust along which blocks of the crust slide relative to one another. Another cross-cutting feature is an intrusion. An *intrusion* is molten rock from the Earth's interior that squeezes into existing rock and cools. **Figure 4** illustrates both of these features.

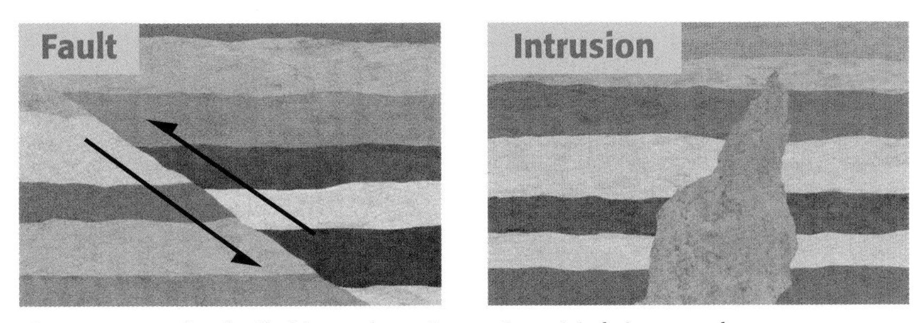

Figure 4 *A fault (left) and an intrusion (right) are always younger than the layers they cut across.*

Geologists assume that the way sediment is deposited to form rock layers—in horizontal layers—has not changed over time. According to this principle, if rock layers are not horizontal, something must have disturbed them after they formed. This principle allows geologists to determine the relative ages of rock layers and the events that disturbed them.

Folding and tilting are two additional types of events that disturb rock layers. *Folding* occurs when rock layers bend and buckle from Earth's internal forces. *Tilting* occurs when internal forces in the Earth slant rock layers without folding them. **Figure 5** illustrates the results of folding and tilting.

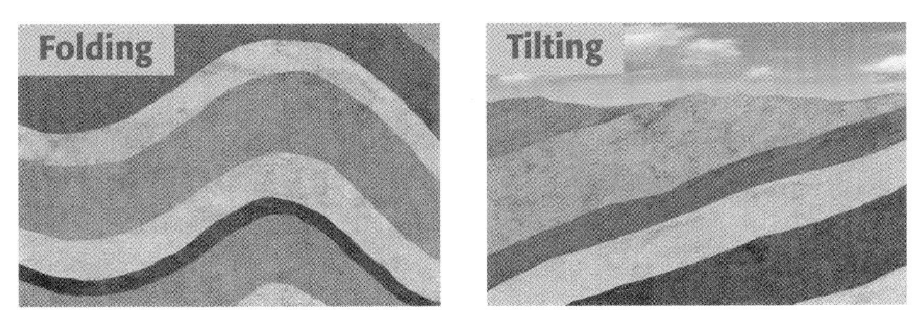

Figure 5 *Folding (left) and tilting (right) are events that are always younger than the rock layers they affect.*

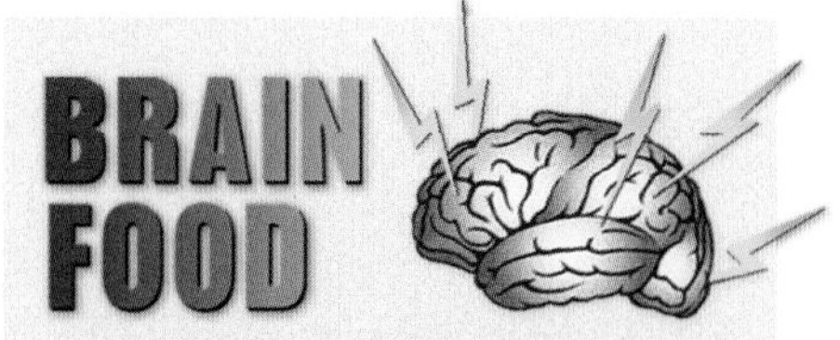

Many high-rise apartment and office buildings exhibit something similar to an unconformity—they do not have a 13th floor. Instead, the floors skip from 12 to 14.

Gaps in the Record—Unconformities

Faults, intrusions, and the effects of folding and tilting can make dating rock layers a challenge. But sometimes layers of rock are missing altogether, creating a gap in the geologic record. To think of this another way, let's say that you stack your newspapers every day after reading them. Now let's suppose you want to look at a paper you read 10 days ago. You know that the paper you want should be 10 papers deep in the stack. But when you look, the paper is not there. What happened? Perhaps you forgot to put the paper in the stack. Now instead of a missing newspaper, imagine a missing rock layer.

Missing Evidence Missing rock layers create gaps in rock-layer sequences called unconformities. An **unconformity** is a surface that represents a missing part of the geologic column. Unconformities also represent missing time—time that was not recorded in layers of rock. When geologists find unconformities, they must question whether the "missing layers" were actually present or whether they were somehow removed. **Figure 6** shows how *nondeposition* and *erosion* create unconformities.

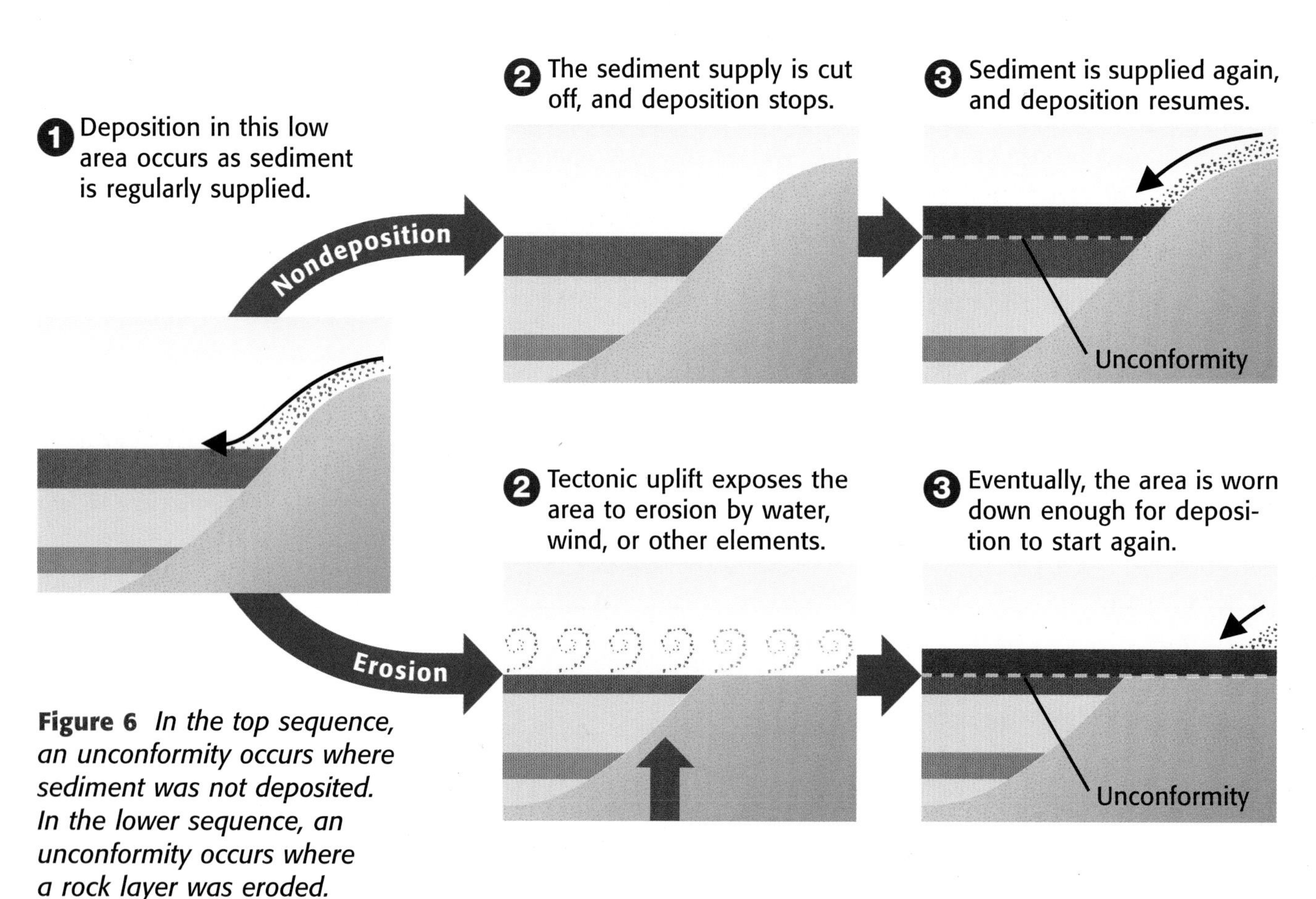

Figure 6 *In the top sequence, an unconformity occurs where sediment was not deposited. In the lower sequence, an unconformity occurs where a rock layer was eroded.*

Types of Unconformities

Most unconformities form by both erosion and nondeposition. But other factors can complicate matters. To simplify the study of unconformities, geologists put them in three major categories—disconformities, nonconformities, and angular unconformities. The three diagrams at right illustrate these three categories.

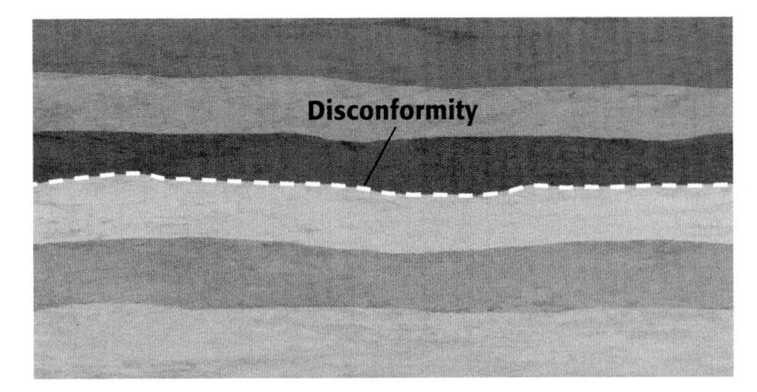

Figure 7 *A* **disconformity** *exists where part of a sequence of parallel rock layers is missing. While often hard to see, a disconformity is the most common type of unconformity.*

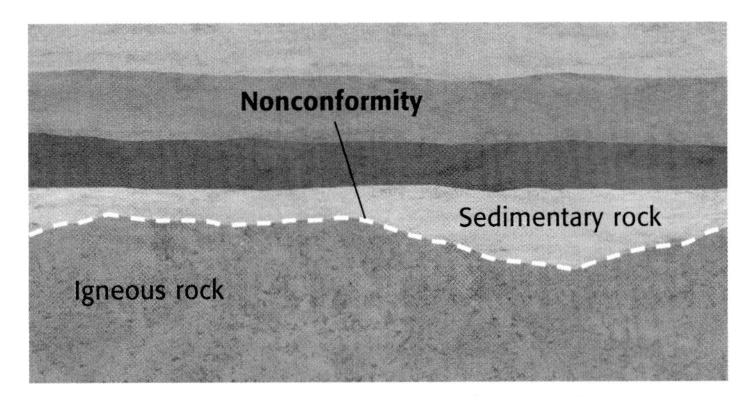

Figure 8 *A* **nonconformity** *exists where sedimentary rock layers lie on top of an eroded surface of non-layered igneous or metamorphic rock.*

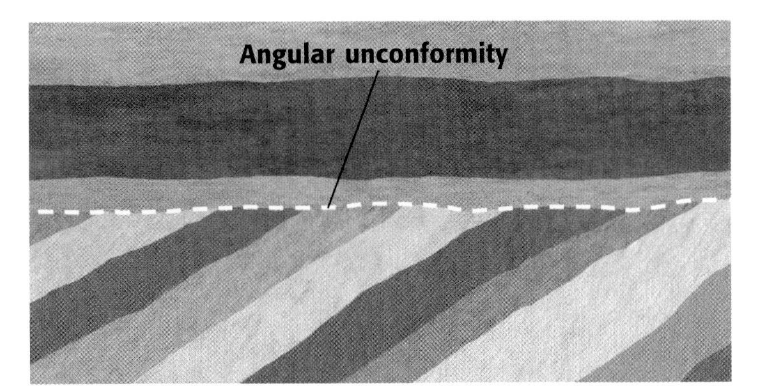

Figure 9 *An* **angular unconformity** *exists between horizontal rock layers and rock layers that are tilted or folded. The tilted or folded layers were eroded before horizontal layers formed above them.*

Rock-Layer Puzzles

Geologists often find rock-layer sequences that have been affected by more than one of the events and features mentioned in this section. For example, an intrusion may squeeze into rock layers that contain an unconformity and that have been cut across by a fault. Determining the order of events that led to such a sequence is like piecing together a jigsaw puzzle.

SECTION REVIEW

1. In a rock-layer sequence that hasn't been disturbed, are older layers found on top of younger layers? What rule do you use to answer this question?
2. List five events or features that can disturb rock-layer sequences.
3. Consider a fault that cuts through all the layers of a rock-layer sequence. Is the fault older or younger than the layers? Explain.
4. **Analyzing Methods** Unlike other types of unconformities, disconformities are hard to recognize because all the layers are horizontal. How does a geologist know when he or she is looking at a disconformity?

SECTION 3

READING WARM-UP

Absolute Dating: A Measure of Time

Terms to Learn

absolute dating
isotopes
radioactive decay
radiometric dating
half-life

What You'll Do

- Explain how radioactive decay occurs.
- Explain how radioactive decay relates to radiometric dating.
- List three types of radiometric dating.
- Determine the best type of radiometric dating to use to date an object.

By using relative dating, scientists can determine the relative ages of rock layers. To determine the actual age of a layer of rock or a fossil, however, scientists must rely on absolute dating. **Absolute dating** is a process of establishing the age of an object, such as a fossil or rock layer, by determining the number of years it has existed. In this section, we will concentrate on radiometric dating, which is the most common method of absolute dating.

Radioactive Decay

To determine the absolute ages of fossils and rocks, scientists most often analyze radioactive isotopes. **Isotopes** are atoms of the same element that have the same number of protons but have different numbers of neutrons. Most isotopes are stable, meaning that they stay in their original form. But some isotopes are unstable. Scientists call unstable isotopes *radioactive*. Radioactive isotopes tend to break down into stable isotopes of other elements in a process called **radioactive decay. Figure 10** shows how one type of radioactive decay occurs. Because radioactive decay occurs at a steady pace, scientists can use the relative amounts of stable and unstable isotopes present in an object to determine the object's age.

Unstable isotope
6 protons, 8 neutrons

Figure 10 *During radioactive decay, an unstable parent isotope breaks down into a stable daughter isotope.*

Radioactive decay
When the unstable isotope decays, a neutron is converted into a proton. In the process, an electron is released.

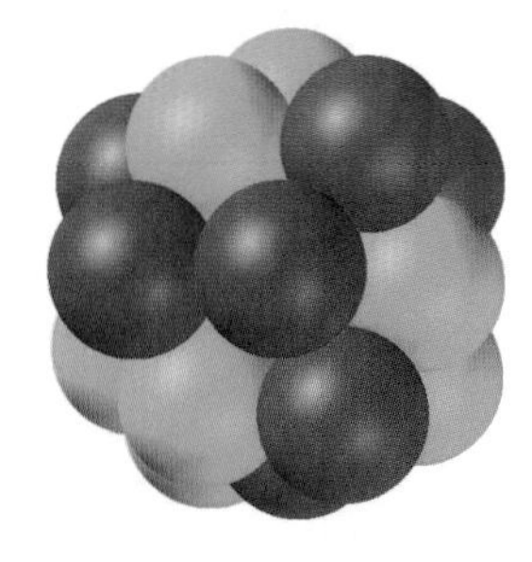

Stable isotope
7 protons, 7 neutrons

Dating Rocks—How Does It Work? Consider a stream of molten lava flowing out of a volcano. As long as the lava is in liquid form, the daughter material that is already present and the parent material are free to mix and move around. But eventually the lava cools and becomes solid igneous rock. When this happens, the parent and daughter materials often end up in different minerals. Scientists know that any daughter material found in the same mineral as the parent material most likely formed after the lava became solid rock. Scientists compare the amount of new daughter material with the amount of parent material that remains. The more new daughter material there is, the older the rock is.

Radiometric Dating

If you know the rate of decay for an element in a rock, you can figure out the age of the rock. Determining the absolute age of a sample based on the ratio of parent material to daughter material is called **radiometric dating.** For example, let's say that it takes 10,000 years for half the parent material in a rock sample to decay into daughter material. You analyze the sample and find equal amounts of parent material and daughter material. This means that half the original radioactive isotope has decayed and that the sample must be about 10,000 years old.

What if one-fourth of your sample is parent material and three-fourths is daughter material? You would know that it took 10,000 years for half the original sample to decay and another 10,000 years for half of what remained to decay. The age of your sample would be $2 \times 10,000$, or 20,000, years. **Figure 11** shows how this steady decay works. The time it takes for one-half of a radioactive sample to decay is called a **half-life.**

MATHBREAK

Get a Half-Life!

After observing the process illustrated in Figure 11, complete the chart below in your ScienceLog.

Parent left	Half-life in years	Age in years
1/8	?	30,000
?	1.3 billion	3.9 billion
1/4	10,000	?

Figure 11 *After every half-life, the amount of parent material decreases by one-half.*

1/1

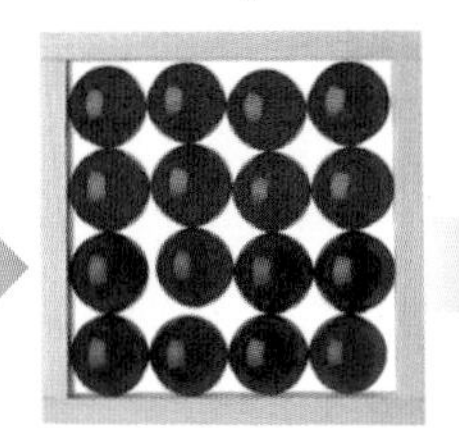

1/2

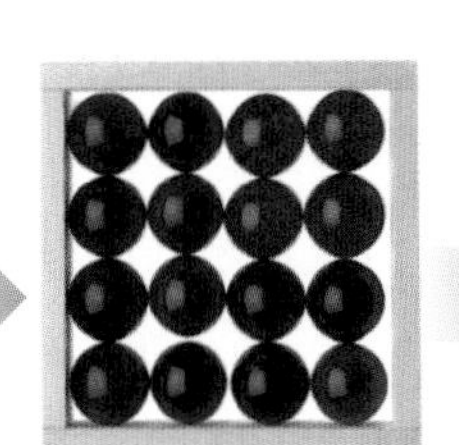

1/4

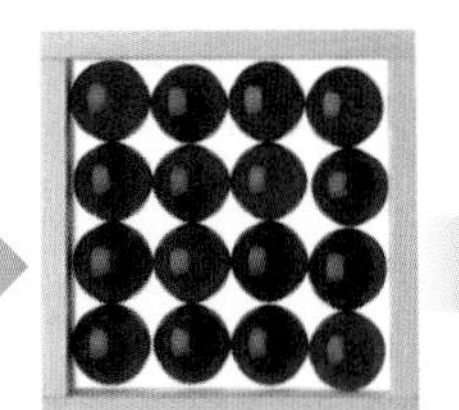

1/8

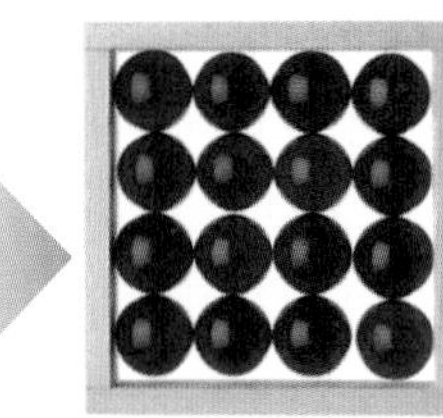

1/16

Types of Radiometric Dating

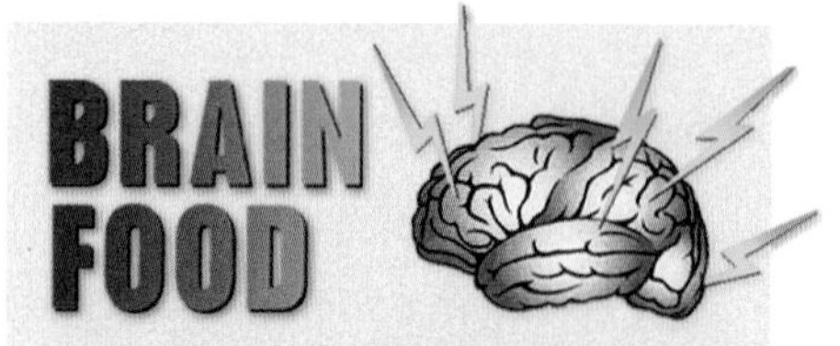

Did you know that scientists have radiometrically dated moon rocks? The ages they have determined suggest that the moon formed when the Earth was still molten.

Imagine traveling back through the centuries to a time long before Columbus arrived in America. You are standing along the bluffs of what will one day be called the Mississippi River. You see dozens of people building large mounds. Who are these people, and what are they building?

The people you saw in your time travel were American Indians, and the structures they were building were burial mounds. The area you imagined is now an archaeological site called Effigy Mounds National Monument. **Figure 12** shows one of these mounds.

According to archaeologists, people lived at Effigy Mounds from 2,500 years ago to 600 years ago. How do archaeologists know these dates? They have dated bones and other objects in the mounds using radiometric dating. Scientists use different radiometric dating techniques based on the estimated age of an object. As you read on, think about how the half-life of an isotope relates to the age of the object being dated. Which technique would you use to date the burial mounds?

Figure 12 *This burial mound at Effigy Mounds resembles a snake.*

Uranium-Lead Method Uranium-238 is a radioactive isotope that eventually decays to lead-206. The half-life of uranium-238 is 4.5 billion years. The older the rock is, the more daughter material (lead-206) there will be in the rock. Uranium-lead dating can be used for rocks more than 10 million years old. Younger rocks do not contain enough daughter material to be accurately measured by this method.

Potassium-Argon Method Another isotope used for radiometric dating is potassium-40. Potassium-40 has a half-life of 1.3 billion years, and it eventually decays to argon and calcium. Geologists measure argon as the daughter material for radiometric dating. This method is mainly used to date rocks older than 100,000 years.

Carbon-14 Method The carbon-14 method works differently from the two methods already mentioned. The element carbon is normally found in three forms, the stable isotopes carbon-12 and carbon-13 and the radioactive isotope carbon-14. These carbon isotopes combine with oxygen to form the gas carbon dioxide, which is taken in by plants during photosynthesis. As long as a plant is alive, new carbon dioxide with a constant carbon-14 to carbon-12 ratio is continually taken in. Animals that eat plants contain the same ratio of carbon isotopes.

Once a plant or animal dies, however, no new carbon is taken in. The amount of carbon-14 begins to decrease as the plant or animal decays, and the ratio of carbon-14 to carbon-12 decreases. This decrease can be measured in a laboratory, such as the one shown in **Figure 13.** Because the half-life of carbon-14 is only 5,730 years, this dating method is mainly used for dating things that lived within the last 50,000 years.

Figure 13 *Some samples containing carbon must be cleaned and burned before their age can be determined.*

SECTION REVIEW

1. Explain how radioactive decay occurs.
2. How does radioactive decay relate to radiometric dating?
3. List three types of radiometric dating.
4. **Applying Concepts** Which radiometric-dating method would be most appropriate for dating artifacts found at Effigy Mounds? Explain.

SECTION 4
READING WARM-UP

Looking at Fossils

Terms to Learn

fossil
permineralization
petrification
trace fossil
coprolite
mold
cast
index fossil

What You'll Do

- Describe how different types of fossils are formed.
- List the types of fossils that are not part of organisms.
- Demonstrate how fossils can be used to determine changes in environments and in the organisms the fossils came from.
- Describe index fossils, and explain how they are used.

Imagine you and your classmates are on a cross-country science field trip to Coralville, a town in east-central Iowa. Your teacher takes your class to a nearby stone quarry and points to a large rock wall that looks just like a coral reef. "This is how Coralville got its name," your teacher explains. "There used to be a living coral reef right here. What you see today is a fossilized coral reef." But you know that coral reefs are found in warm tropical oceans and that Iowa is more than 1,000 km away from any ocean! How did this huge coral reef end up in the middle of Iowa? To answer this question, you need to learn about fossils.

Fossilized Organisms

A **fossil** is any naturally preserved evidence of life. Fossils exist in many forms. The most easily recognizable fossils are preserved organisms, such as the stingray shown at right, or parts of organisms. Usually these fossils occur in rock. But as you will see, other materials can also preserve evidence of life.

Fossils in Rocks When organisms die, the soft, fleshy parts of their bodies decompose, leaving only the hard parts. Occasionally, these hard parts get buried quickly in sediment and are preserved while the sediment turns to rock.

It takes more time for hard body parts such as bones, shells, and wood to decompose. For this reason, organisms with hard body parts are more likely to become fossils than those with only soft parts.

Mineral Replacement Organisms can also be preserved by **permineralization,** a process in which minerals fill in pore spaces of an organism's tissues. Minerals can also replace the original tissues of organisms. **Petrification** of an organism, shown in **Figure 14,** occurs when the organism's tissues are completely replaced by minerals.

Figure 14 *These pieces of petrified wood are made of stone.*

Fossils in Amber Imagine a fly or a mosquito landing in a drop of tree sap and getting stuck. Suppose that the insect gets covered by more sap. When the sap hardens, the insect will be preserved inside. Hardened tree sap is called *amber.* Some of our best insect fossils are found in amber, as shown in **Figure 15.**

Figure 15 *This insect is perfectly preserved in amber.*

Mummification When organisms die in dry places, such as deserts, they can sometimes dry out so fast that there isn't enough time for even their soft parts to decay. This process is called *mummification.* Mummified organisms don't decay because the bacteria that feed on dead organisms can't live without water. Some food, like dried fruit and beef jerky, is preserved in a similar way.

Frozen Fossils Imagine a huge animal that looks like an elephant with long hair walking along a glacier 12,000 years ago. It's a woolly mammoth. Suddenly, the beast slips and falls between two huge pieces of ice into a deep crack. With no way out, the animal freezes and is preserved until the glacier thaws thousands of years later. Scientists find fossils of woolly mammoths and many other organisms when glaciers thaw. These frozen specimens are some of the best fossils.

Fossils in Tar There are places where tar occurs naturally in thick, sticky pools. One such place is the La Brea tar pits, in Los Angeles County, California. These pits of thick oil and tar were present when saber-toothed cats roamed the Earth 40,000 years ago, as shown in **Figure 16.**

Much of what we know about these extinct cats comes from fossils found in the La Brea tar pits. But saber-toothed cats are not the only organisms found in the pits. Scientists have found fossils of many other mammals as well as plants, snails, birds, salamanders, and insects.

Figure 16 *Many animals, including saber-toothed cats, became fossils after sinking in tar pits.*

Other Types of Fossils

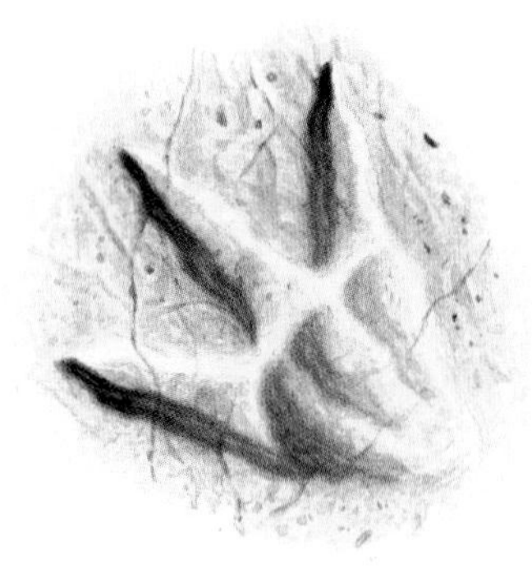

What happens when scientists cannot find any remains of plants or animals? Is there anything else that might indicate an organism's former presence?

Trace Fossils Any naturally preserved evidence of an animal's activity is called a **trace fossil.** An easily recognizable type of trace fossil is a *track*. Just like animals today, the animals in the past left tracks. These ancient tracks became fossils when they filled with sediment that eventually turned to rock.

Imagine that the tracks shown here were made by a ferocious *Tyrannosaurus rex*. While the animal that made them is long gone, the fossil tracks remain as evidence that it once prowled the Earth.

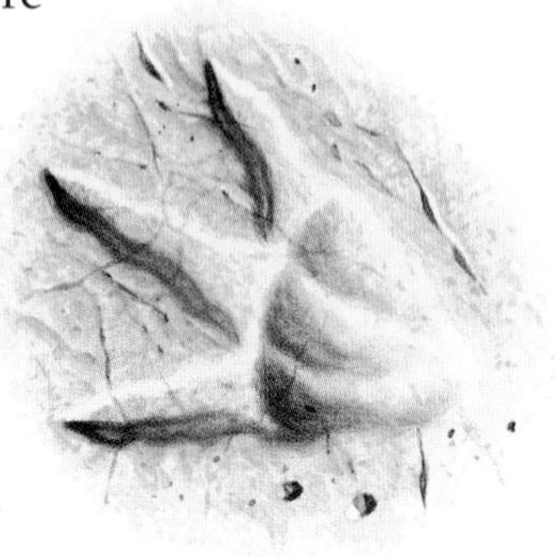

Burrows are another type of trace fossil. Burrows are shelters made by animals that dig into the ground. Like tracks, burrows are preserved when they are filled in with sediment and buried quickly.

Coprolites are a third type of trace fossil. The word *coprolite* (KAHP roh LIET) is from the Greek words meaning "dung stone." **Coprolites** are preserved feces, or dung, from animals. Coprolites can provide valuable information about the habits and diets of the animals that left them. **Figure 17** shows a coprolite that is more than 5 million years old.

Figure 17 *This coprolite came from a prehistoric mammal.*

Self-Check

Why are tracks and coprolites considered trace fossils? *(See page 216 to check your answer.)*

Molds and Casts A **mold** is a cavity in the ground or rock where a plant or animal was buried. Often the cavity has been filled in, leaving a cast of the original organism. A **cast** is an object created when sediment fills a mold and becomes rock. A cast shows what the outside of the organism looked like. **Figure 18** shows a mold and cast from the same organism.

Figure 18 *The ammonite cast on the left formed when sediment filled the ammonite mold on the right and became rock.*

Using Fossils to Interpret the Past

By examining fossils, scientists can find out what was happening in the environment when the sediments surrounding the fossils were deposited. Scientists can also interpret how plants and animals have changed over time by studying fossils from different parts of the geologic column.

Changes in Environments

Fossils can reveal changes that have occurred in parts of the Earth. By studying the coral-reef fossils and applying the principle of uniformitarianism, for example, scientists have determined that Iowa was once covered by a shallow sea. This is hard to believe when you look at Iowa's landscape today!

Iowa is just one example of where inconsistent fossils have been found. Who would have expected fossils of coral to be found in the landlocked state of Iowa? Likewise, who would have expected fossils of marine organisms on the top of a mountain? But that is exactly what scientists found on mountaintops in Canada, as shown in **Figure 19.** The presence of these fossils means that these rocks were once below the surface of an ocean.

Figure 19 *Scientists often find rocks that contain marine fossils on mountaintops. These rocks were pushed up from below sea level millions of years ago.*

Changes in Life Older rock layers contain organisms different from those found in younger rock layers. The record stored in the rocks shows a change in life-forms over the years. For example, rock layers that contain fish fossils are found beneath the oldest rock layers that contain fossils of amphibians. Amphibians, such as frogs and salamanders, are animals with characteristics that allow them to live both on land and in water. On top of these rock layers are the oldest layers that contain fossils of reptiles, most of which lived only on land. Using the principle of superposition, we know that fish existed before amphibians because fish were found in a lower layer of rock. In the same way, we know that amphibians existed before reptiles.

Using Fossils to Date Rocks

Geologists sometimes use *index fossils* to date rocks while in the field. **Index fossils** are fossils of organisms that lived during a relatively short, well-defined time span. Whenever geologists find an index fossil in a rock layer, they know where in the geologic column the rock layer fits. This enables them to give the layer a date without directly using radiometric dating. Good index fossils also have a wide distribution around the world.

An example of an index fossil is a genus of trilobites called *Phacops,* shown above. Trilobites are extinct, but they looked like a cross between a modern horseshoe crab and a pill bug. *Phacops* lived in shallow oceans about 400 million years ago. Where geologists find a fossil of this trilobite, they can assume that the surrounding rock is about 400 million years old.

Figure 20 Tropites, *a genus of ammonites, existed for only about 20 million years, which makes it a good index fossil.*

Another good index fossil is a genus of ammonites called *Tropites,* shown in **Figure 20.** Ammonites were marine animals that looked a lot like modern squids, but they lived in coiled shells with complex inner walls. *Tropites* lived between 230 million and 208 million years ago. Where geologists find them in a rock layer, they know that the rock layer is between 208 million and 230 million years old.

internet**connect**

sciLINKS NSTA

TOPIC: Looking at Fossils
GO TO: www.scilinks.org
***sci*LINKS NUMBER:** HSTE145

SECTION REVIEW

1. Describe two ways that fossils can form.
2. List two types of fossils that are not part of an organism.
3. What are index fossils? How do scientists use them to date rocks?
4. **Making Inferences** If you find rock layers containing fish fossils in a desert, what can you infer about that area of the desert?

SECTION 5

READING WARM-UP

Time Marches On

Terms to Learn

geologic time scale
eon
period
era
epoch

What You'll Do

- Demonstrate an understanding of the geologic time scale.
- Identify important dates on the geologic time scale.
- Identify the eon we know the most about, and explain why we know more about it than about other eons.

Remember the stack of family pictures mentioned in Section 2? The oldest pictures were on the bottom, and the newest ones were on the top. By looking through the pictures in order, you could see the sequence of events and changes that occurred in your family's history. In studying the history of the Earth, scientists follow a similar process. But instead of looking at pictures, they analyze rock layers and the fossils they contain.

Rock Layers and Geologic Time

One of the best places in North America to see the Earth's history recorded in rock layers is in Grand Canyon National Park, shown in **Figure 21.** The Colorado River has cut the canyon nearly 2 km deep in some places. During this process, countless layers of rock have been eroded by the river. These layers represent nearly 2 billion years of geologic time!

Figure 21 *The rock layers in the Grand Canyon correspond to a very large section of the geologic column.*

Biology CONNECTION

The Grand Canyon is so wide and deep that organisms on either side of the canyon took different evolutionary paths. As the Colorado River formed the canyon, groups of individuals from the same species became separated and could no longer interact. Over millions of years, these groups developed differently and became different species.

The Geologic Time Scale

While the rock layers in the Grand Canyon represent the time that passed as they formed, the geologic column represents the billions of years that have passed since the first rocks formed on Earth. Geologists must grapple with the time represented by the geologic column as well as the time between Earth's formation and the formation of Earth's oldest known rocks. Altogether, geologists study 4.6 billion years of Earth's history! To make their job easier, geologists have created the geologic time scale. The **geologic time scale,** which is shown in **Figure 22,** is a scale that divides Earth's 4.6-billion-year history into distinct intervals of time.

Figure 22 *The geologic time scale accounts for Earth's entire history. It is divided into four major parts called* eons.

Geologic Time Scale

Eon	Era	Period	Epoch	Millions of years ago
PHANEROZOIC EON	Cenozoic	Quaternary	Holocene	0.01
			Pleistocene	1.8
		Tertiary	Pliocene	5.3
			Miocene	23.8
			Oligocene	33.7
			Eocene	54.8
			Paleocene	65
	Mesozoic	Cretaceous		144
		Jurassic		206
		Triassic		248
	Paleozoic	Permian		290
		Pennsylvanian		323
		Mississippian		354
		Devonian		417
		Silurian		443
		Ordovician		490
		Cambrian		540
PROTEROZOIC EON				2,500
ARCHEAN EON				3,800
HADEAN EON				4,600

Phanerozoic eon

(540 million years ago–present)
The rock and fossil record mainly represents the Phanerozoic eon, which is the eon in which we live.

Proterozoic eon

(2.5 billion years ago–540 million years ago)
The first organisms with well-developed cells appeared during this eon.

Archean eon

(3.8 billion years ago–2.5 billion years ago)
The earliest known rocks on Earth formed during this eon.

Hadean eon

(4.6 billion years ago–3.8 billion years ago)
The only rocks that scientists have found from this eon are meteorites and rocks from the moon.

Divisions of Time Geologists have divided Earth's history into sections of time, as shown on the geologic time scale in Figure 22. The largest divisions of geologic time are **eons.** The four eons in turn are divided into **eras,** which are the second-largest divisions of geologic time. Eras are divided into **periods,** which are the third-largest divisions of geologic time. Some periods are divided into **epochs** (EP uhks), which are the fourth-largest division of geologic time. Look again at Figure 22. Can you figure out what epoch we live in?

The boundaries between geologic time intervals represent major changes on Earth. These changes include the appearance or disappearance of life-forms, changes in the global climate, and changes in rock types. For example, each of the three eras of the Phanerozoic eon are characterized by unique life-forms.

Living in the Past
How do scientists know what life was like in prehistoric times? Turn to page 85 to learn how one paleontologist finds out.

The Paleozoic Era *Paleozoic* means "old life." The Paleozoic era lasted from about 540 to 248 million years ago. It is the first era that is well represented by fossils.

At the beginning of the Paleozoic era, there were no land organisms. Imagine how empty the landscape must have looked! By the middle of the era, plants started appearing on land. By the end of the era, amphibians were living partially on the land, and insects were abundant. **Figure 23** shows what the land might have looked like late in the Paleozoic era. The Paleozoic era came to an end with a mass extinction—nearly 90 percent of all species perished.

Figure 23 *Jungles were present during the Paleozoic era, but there were no birds singing in the trees and no monkeys swinging from the branches. Birds and mammals didn't evolve until much later.*

The Mesozoic Era *Mesozoic* means "middle life." The Mesozoic era lasted from about 248 million years ago until about 65 million years ago. This era is also known as the Age of Reptiles. Dinosaurs, such as the ones shown in **Figure 24,** inhabited the land and the water.

Although reptiles dominated the Mesozoic era, birds and small mammals began to evolve late in the era. Most scientists think that birds evolved directly from a type of dinosaur. By the end of the Mesozoic era, about 50 percent of all species on Earth, including the dinosaurs, became extinct.

Figure 24 *Imagine walking in the desert and bumping into these fierce creatures! It's a good thing humans didn't evolve in the Mesozoic era, which was dominated by dinosaurs.*

The Cenozoic Era *Cenozoic* means "recent life." The Cenozoic era began about 65 million years ago and continues to the present. We live in the Cenozoic era.

Whereas the Mesozoic era is called the Age of Reptiles, the Cenozoic era is called the Age of Mammals. After the mass extinction at the end of the Mesozoic era, mammals became abundant on Earth, as shown in **Figure 25.** Many types of mammals that lived earlier in the Cenozoic era are now extinct, including woolly mammoths, saber-toothed cats, and giant sloths.

Figure 25 *Thousands of species of mammals evolved during the Cenozoic era. This scene shows species from the early Cenozoic era that are now extinct.*

Can You Imagine 4.6 Billion Years?

It's hard to picture 4.6 billion of anything, especially years. As humans, we do quite well to live to be 100 years old. Given this perspective, it is very difficult to think of Earth as being billions of years old. One way to do this is to organize the geologic time scale into the frame of 12 hours, with the first moment of Earth's history being noon and the present moment being midnight. This has been done on the Earth-history clock shown in **Figure 26.** On the Earth-history clock, the millions of years of evolution that you just read about occurred within the last hour. Human civilizations appeared within the last second! Perhaps you now have a better understanding of just how old the Earth is and just how brief humans' existence has been.

Make a Time Scale

1. Using a pair of **scissors,** cut a length of **adding-machine tape** 46 cm long.
2. Starting at one end of the tape, use a **ruler** and a **black marker** to draw a line across the width of the tape at the following measurements: 5.4 cm, 25 cm, and 38 cm.
3. Using **colored markers,** color the sections of tape as follows:
   0 cm–5.4 cm = green
   5.4 cm–25 cm = blue
   25 cm–38 cm = red
   38 cm–46 cm = yellow
4. Your tape represents the geologic time scale, and the present moment is at 46 cm. What is the name of each time interval on your scale?

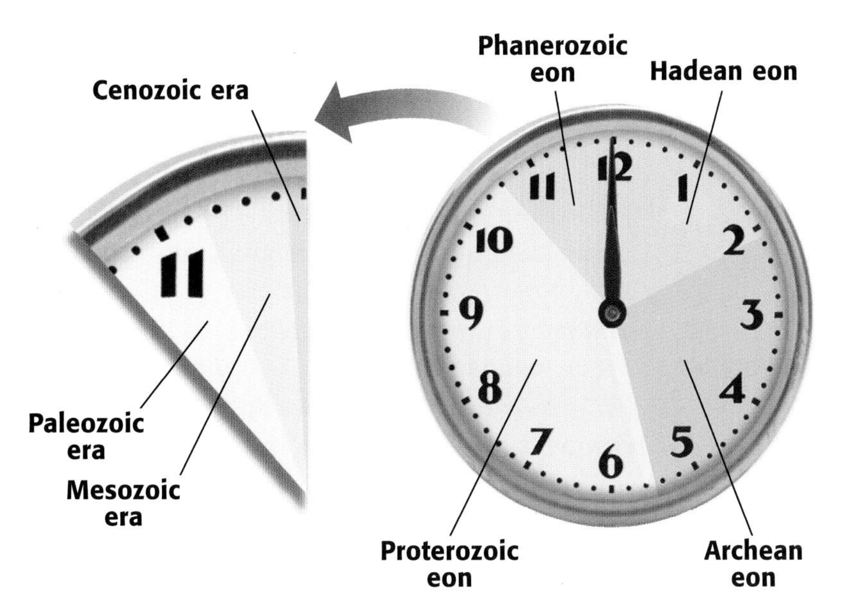

Figure 26 *On the Earth-history clock, which organizes Earth's history into the frame of 12 hours, 1 hour equals 383 million years, 1 minute equals 6.4 million years, and 1 second equals 106,000 years.*

SECTION REVIEW

1. How many eras are in the Phanerozoic eon? List them.
2. In this section, extinctions at the end of two geologic time intervals are mentioned. What are these two intervals, and when did each interval end?
3. Which eon do we know the most about? Why?
4. **Making Predictions** What future event might mark the end of the Cenozoic era?

internet**connect**

TOPIC: Geologic Time
GO TO: www.scilinks.org
***sci*LINKS NUMBER:** HSTE150

Discovery Lab

How DO You Stack Up?

According to the *principle of superposition,* in undisturbed sequences of sedimentary rock, the oldest layers are on the bottom. Geologists use this principle to determine the relative age of the rocks in a small area. In this activity, you will model what geologists do by drawing sections of different rock outcrops. Then you will create a part of the geologic column, showing the geologic history of the area that contains all of the outcrops.

MATERIALS

- metric ruler
- pencil
- colored pencils or crayons
- white paper
- scissors
- transparent tape

Procedure

1. Use a metric ruler and a pencil to draw four boxes on a blank sheet of paper. Each box should be 3 cm wide and at least 6 cm tall. (You can trace the boxes shown on the next page.)
2. With colored pencils, copy the illustrations of the four outcrops on the next page. Use colors and patterns similar to those shown.
3. Pay close attention to the contact between layers—straight or wavy. Straight lines represent bedding planes, where deposition was continuous. Wavy lines represent unconformities, where rock layers may be missing. The top of each outcrop is incomplete, so it should be a jagged line. (Assume that the bottom of the lowest layer is a bedding plane.)
4. Use a black crayon or pencil to add the symbols representing fossils to the layers in your drawings. Pay attention to the variety of fossil shapes and the layers that they are in.
5. Write the outcrop number on the back of each section.
6. Carefully cut the outcrops out of the paper, and lay the individual outcrops next to each other on your desk or table.
7. Find layers that have the same rocks and contain the same fossils. Move each outcrop up or down to align similar layers next to each other.

8. If unconformities appear in any of the outcrops, rock layers may be missing. You may need to examine other sections to find out what fits between the layers above and below the unconformities. Leave room for these layers by cutting the outcrops along the unconformities (wavy lines).

9. Eventually, you should be able to make a geologic column that represents all four of the outcrops. It will show rock types and fossils for all the known layers in the area.

10. Tape the pieces of paper together in a pattern that represents the complete geologic column.

Analysis

11. How many layers are in this part of the geologic column you modeled?

12. Which is the oldest layer in your column? Which rock layer is the youngest? Describe these layers in terms of rock type and the fossils they contain.

13. Which, if any, fossils can be used as index fossils for a single layer? Why are these fossils considered index fossils?

14. List the fossils in your column from oldest to youngest. Label the oldest and youngest fossils.

15. Look at the unconformity in Outcrop 2. Which rock layers are partially or completely missing? Explain how you know this.

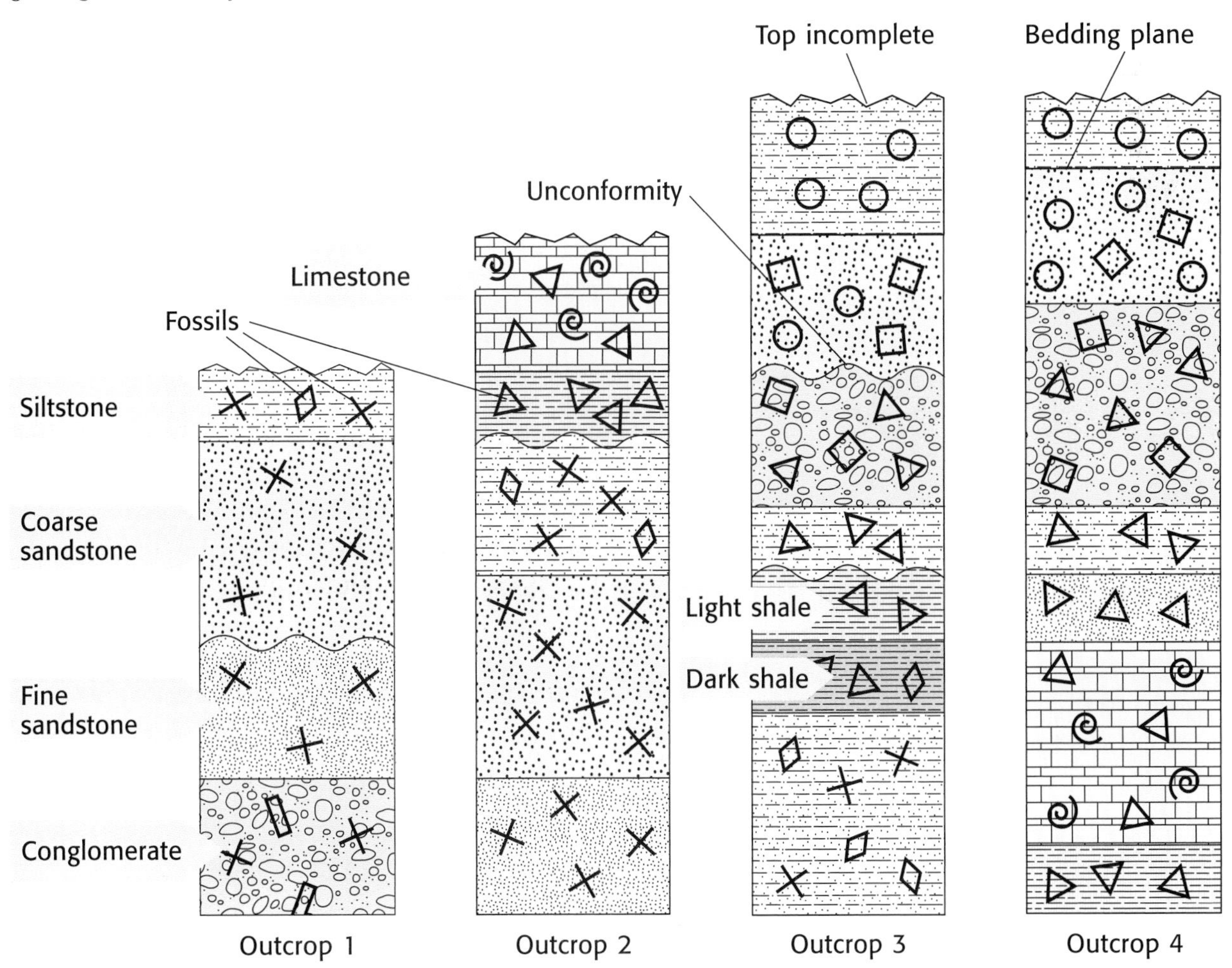

Chapter Highlights

SECTION 1

Vocabulary

uniformitarianism *(p. 56)*
catastrophism *(p. 57)*

Section Notes

- Scientists use the principle of uniformitarianism to interpret the past and make predictions.
- According to uniformitarianism, geologic change is gradual. According to catastrophism, geologic change is sudden.
- Before Hutton and Lyell, most scientists believed all geologic change was catastrophic. After Hutton and Lyell, most scientists rejected catastrophism. Today most scientists favor uniformitarianism, but they recognize some geologic change as catastrophic.

SECTION 2

Vocabulary

relative dating *(p. 59)*
superposition *(p. 59)*
geologic column *(p. 60)*
unconformity *(p. 62)*

Section Notes

- Geologists use relative dating to determine the relative age of objects.
- Geologists assume that younger layers lie above older layers in undisturbed rock-layer sequences. This is called superposition.
- The entire rock and fossil record is represented by the geologic column.
- Geologists examine the relationships between rock layers and the structures that cut across them in order to determine relative ages.
- Geologists also determine relative ages by assuming that all rock layers were originally horizontal.
- Unconformities form where rock layers are missing, and they represent time that is not recorded in the rock record.

Skills Check

Math Concepts

HALF-LIVES Remember from Figure 11 on page 65 that the ratio of parent material to daughter material decreases by one-half with each half-life. An easy way to think of this is to multiply the ratio by 1/2 for each half-life. This is shown below.

$$\frac{1}{1} \times \frac{1}{2} = \frac{1}{2};\ \frac{1}{2} \times \frac{1}{2} = \frac{1}{4};$$

$$\frac{1}{4} \times \frac{1}{2} = \frac{1}{8};\ \text{and}\ \frac{1}{8} \times \frac{1}{2} = \frac{1}{16}$$

Visual Understanding

FAULTS AND UNCONFORMITIES It is important to realize that faults and unconformities are not bodies of rock. They are types of surfaces where bodies of rock contact each other.

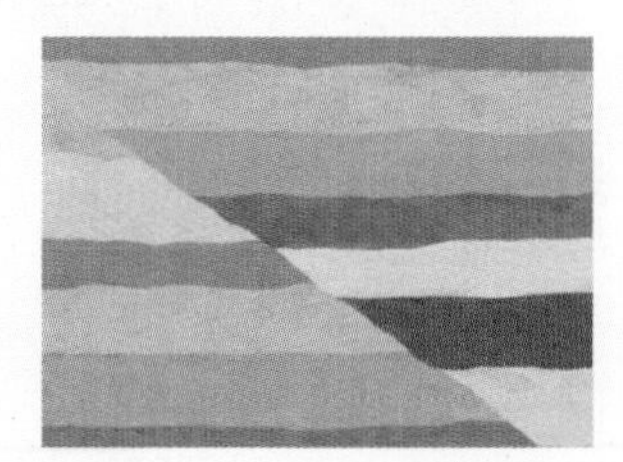

Concept Mapping

17. Use the following terms to create a concept map: age, absolute dating, half-life, radioactive decay, radiometric dating, relative dating, superposition, geologic column, isotopes.

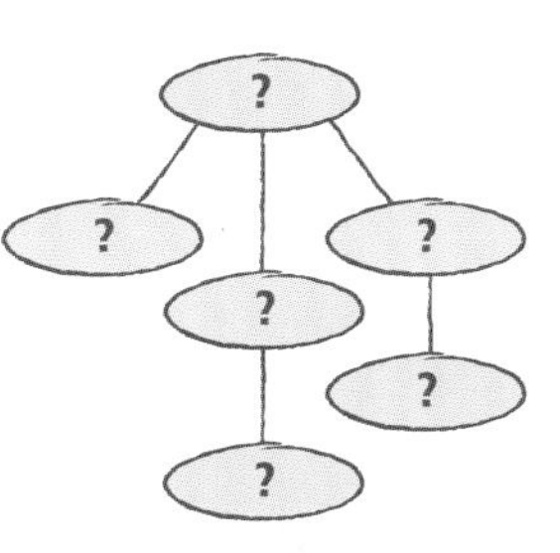

CRITICAL THINKING AND PROBLEM SOLVING

Write one or two sentences to answer the following questions:

18. You may have heard the term *petrified wood.* Why doesn't a "petrified" tree contain any wood?
19. How do tracks and burrows end up in the rock and fossil record?
20. How do you know that an intrusion is younger than its surrounding rock layers?

MATH IN SCIENCE

21. Copy the graph below onto a separate sheet of paper. Place a dot on the *y*-axis at 100 percent. Then place a dot on the graph at each half-life to show how much of the parent material is left. Connect the points with a curved line. Will the percentage of parent material ever reach zero? Explain.

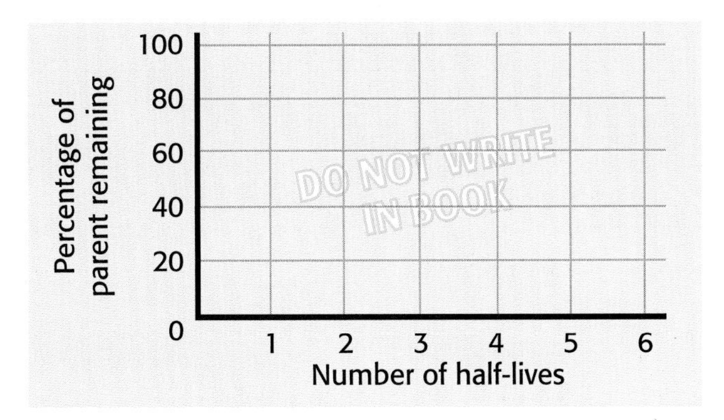

INTERPRETING GRAPHICS

Examine the drawing below, and answer the following questions.

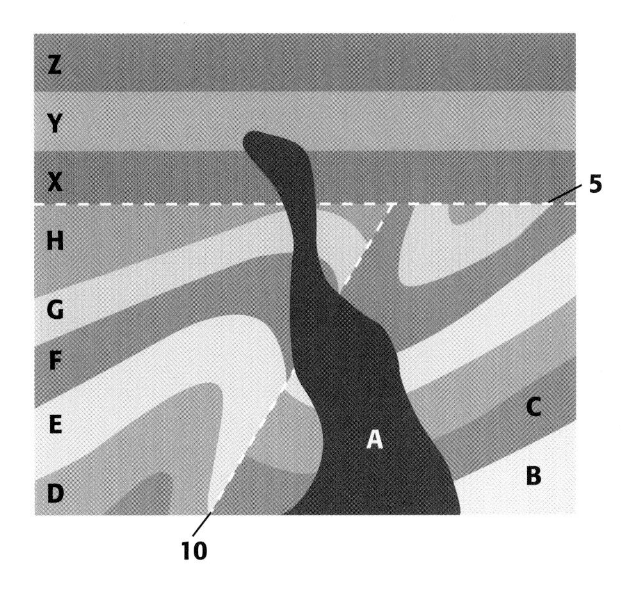

22. Is intrusion **A** younger or older than layer **X**?
23. What kind of unconformity is marked by **5**?
24. Is intrusion **A** younger or older than fault **10**? Why?
25. Other than the intrusion and faulting, what event occurred in layers **B**, **C**, **D**, **E**, **F**, **G**, and **H**? Number this event, the intrusion, and the faulting in the order they occurred.

Reading Check-up

Take a minute to review your answers to the Pre-Reading Questions found at the bottom of page 54. Have your answers changed? If necessary, revise your answers based on what you have learned since you began this chapter.

Science, Technology, and Society

CAT Scanning Fossils

Imagine that you've just found the fossilized skull of a small prehistoric mammal. You examine it very carefully, taking note of its size, shape, and external features. But you also want to look at features inside the skull, like the tiny bones of the middle ear. Can you do it without damaging the fossil? In the past it would have been impossible. Today, though, scientists are using medical technology to do this kind of detailed examination.

Breaking Bones

Paleontologists want to learn all they can about the fossils they study. They want to know about internal structures as well as external ones. Paleontologists usually grind a fossil away layer by layer, recording their observations as they go. Unfortunately, by the time they finish analyzing all the internal structures, the fossil is destroyed! This is a real problem if you want to show someone else your discovery.

Scientists with X-ray Vision

Now paleontologists have another choice. *Computerized axial tomography* (CAT scanning) is quickly replacing the more destructive method of studying internal structures. Originally designed as medical technology to examine the inside of the human skull, CAT scans provide interior views of a fossil without even touching its surface.

To understand how a CAT scan works, imagine a dolphin jumping through a hoop. As the dolphin passes through the hoop a CAT scan machine takes an X-ray picture of it from *every point around the hoop.* In effect, the machine takes a series of cross-section X-ray pictures of the dolphin. A computer then assembles these "slices" to create a three-dimensional picture of the dolphin. Every part of the dolphin's insides can then be studied without dissecting the dolphin.

When a paleontologist needs to reconstruct an entire skull, a series of two-dimensional "slice" shots is taken and the "slices" are combined through computer imaging to produce a three-dimensional image of the skull—inside and out!

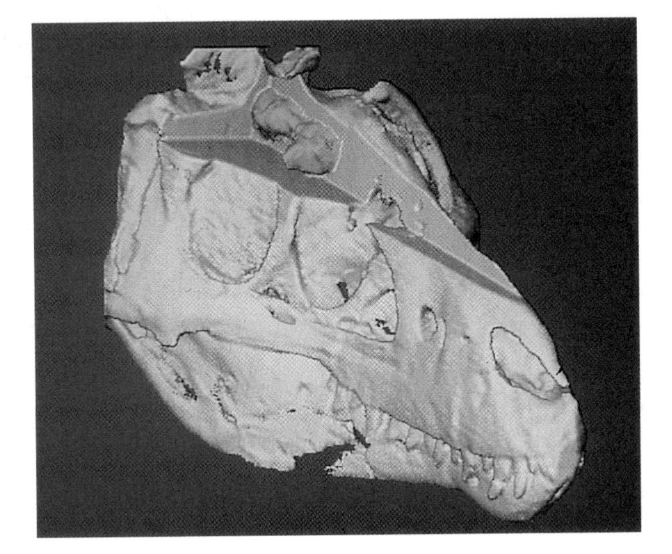

▲ *This CAT scan shows the size and location of the dinosaur* Nanotyrannosaurus rex's *brain.*

What's Hidden Inside?

Using CAT scans, scientists have learned much more about the internal structures of fossils. They have used CAT scans to look at the skeletons of embryos inside fossilized eggs and to study fragile bones still embedded in rock.

On Your Own

▶ What are the advantages of using CAT scans over conventional X rays? Find out by doing some research on your own.

CAREERS

PALEONTOLOGIST

Jack Horner found his first fossil bones at age 7 or 8 while collecting rocks at his father's quarry. From then on, he was hooked on dinosaurs. "I became a paleontologist because I like to dig in the dirt, discover things, and piece together puzzles," Horner says. As one of the world's leading experts on dinosaurs, Horner is curator of paleontology at the Museum of the Rockies, in Bozeman, Montana.

A mother nuzzles her babies in a nest. Nearby, another mother lets out a worried yelp; one of her babies has crawled out of its nest and is scampering away. The mother quickly captures her baby and returns it to safety. Puppies? Birds? No—dinosaurs! Or so Jack Horner believes.

Horner has come to this conclusion by comparing dinosaur fossils with modern alligators and birds. "I am studying how dinosaur bones developed, and I'm comparing them with the development of bones of alligators and birds so that we can learn more about dinosaur growth and nesting behaviors," Horner says. "I think that birds probably evolved from dinosaurs. If I find fossils of several nests close to each other, that tells me that the dinosaurs that built those nests may have lived together in a group."

Meeting the Challenge

As a child, Horner had difficulties in school because he had a learning disability called dyslexia. But no learning disability could dampen Horner's enthusiasm for science, especially the study of dinosaurs. "I like dinosaurs and figuring out what the world looked like at different times in the past. I've always liked the detective work that's involved in paleontology. You can't study a living dinosaur, so you have to figure out everything using clues from the past."

Boning Up on the Latest . . .

One of Horner's current projects is analyzing whether *Tyrannosaurus rex* was a vicious predator, as is often pictured, or a scavenger, eating other animals' kills. The more he studies fossil clues, the more Horner leans toward accepting the scavenger hypothesis. "Predatory animals require certain characteristics in order to be efficient killers. They need to be able to run fast, and they need to be able to maneuver and leap," Horner explains. "*T. rex* couldn't run fast, wasn't agile, and couldn't jump around or even fall down without doing serious damage to itself or even dying."

Decide for Yourself

▶ Observe the behavior of birds in your area. Focus on one or two species. Note their eating habits, the sounds they make, and their interactions with other birds. Do you think birds might have evolved from dinosaurs? Use your observations to support your theory.

▲ *A model of a* Maisasaura *hatching.*

CHAPTER 4

Plate Tectonics

Sections

1. Why do entire mountain ranges move?
2. How do mountains form?

When Continents Collide

The Himalayas are the highest mountains on Earth. They are located between India and Asia in a region where two continents are slowly crashing into each other. This photo shows the highest mountain of all—Mount Everest. At an elevation of 8,848 m, the air at the top of Mount Everest is so thin that climbers must bring their own oxygen! In this chapter you will learn about how and where different types of mountains form. You will also learn about how scientists came up with *plate tectonics,* the theory that revolutionized geology.

Mountain climbers must brave extreme conditions when climbing mountains such as Mount Everest.

CONTINENTAL COLLISIONS

As you can see, continents not only move, but they can also crash into each other. In this activity, you will model the collision of two continents.

Procedure

1. Obtain **two stacks of paper,** each about 1 cm thick.
2. Place the two stacks of paper on a **flat surface,** such as a desk.
3. Very slowly, push the stacks of paper together so that they collide. Continue to push the stacks until the paper in one of the stacks folds over.
4. Repeat step 3, but this time push the two stacks together at a different angle. For example, if you pushed the flat edges together in step 3, try pushing the corners of the paper together this time.

Analysis

5. What happens to the stacks of paper when they collide with each other?
6. Do all of the pieces of paper get pushed upward? If not, what happens to those pieces that do not get pushed upward?
7. What type of landform does this model predict as the result of a continental collision?

SECTION 1

READING WARM-UP

Terms to Learn

- crust
- mantle
- core
- lithosphere
- asthenosphere
- mesosphere
- outer core
- inner core
- tectonic plate

What You'll Do

- Identify and describe the layers of the Earth by what they are made of.
- Identify and describe the layers of the Earth by their physical properties.
- Define *tectonic plate.*
- Explain how scientists know about the structure of Earth's interior.

Inside the Earth

The Earth is not just a ball of solid rock. It is made of several layers with different physical properties and compositions. As you will discover, scientists think about the Earth's layers in two ways—by their *composition* and by their *physical properties.*

Earth's layers are made of different mixtures of elements. This is what is meant by differences in composition. Many of the Earth's layers also have different physical properties. Physical properties include temperature, density, and ability to flow. Let's first take a look at the composition of the Earth.

The Composition of the Earth

The Earth is divided into three layers—the *crust, mantle,* and *core*—based on what each one is made of. The lightest materials make up the outermost layer, and the densest materials make up the inner layers. This is because lighter materials tend to float up, while heavier materials sink.

The Crust The **crust** is the outermost layer of the Earth. Ranging from 5 to 100 km thick, it is also the thinnest layer of the Earth. And because it is the layer we live on, we know more about this layer than we know about the other two.

There are two types of crust—continental and oceanic. *Continental crust* has a composition similar to granite. It has an average thickness of 30 km. *Oceanic crust* has a composition similar to basalt. It is generally between 5 and 8 km thick. Because basalt is denser than granite, oceanic crust is denser than continental crust.

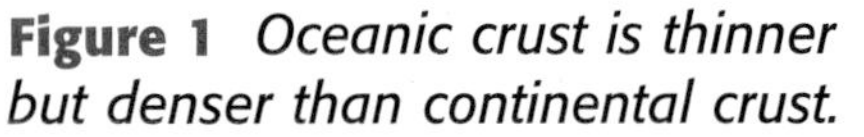

Figure 1 *Oceanic crust is thinner but denser than continental crust.*

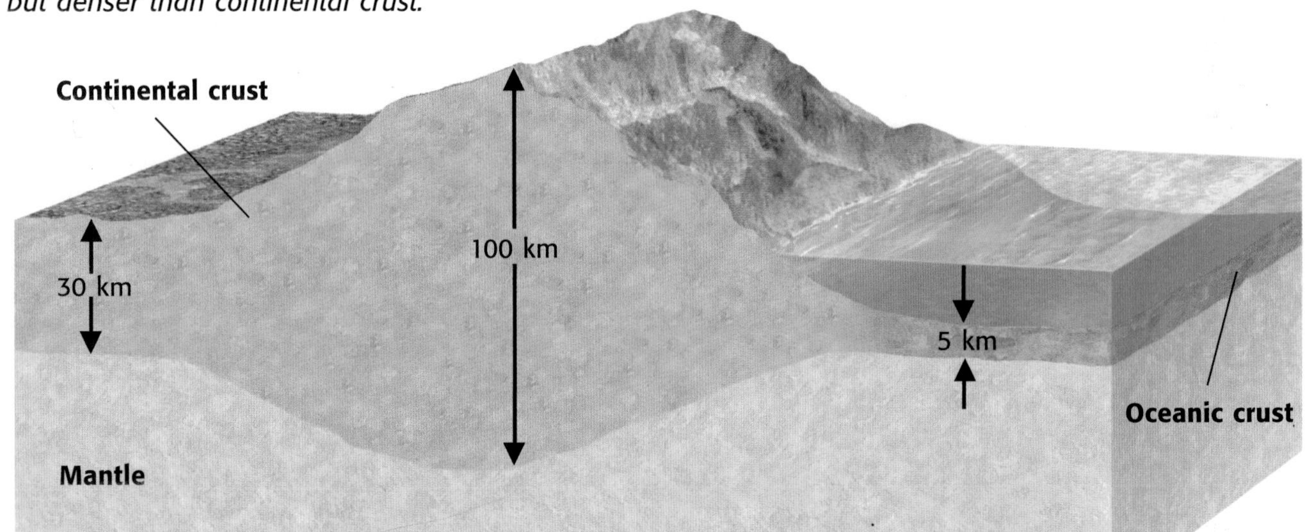

The Mantle The **mantle** is the layer of the Earth between the crust and the core. Compared with the crust, the mantle is extremely thick and contains most of the Earth's mass.

No one has ever seen what the mantle really looks like. It is just too far down to drill for a sample. Scientists must infer what the composition and other characteristics of the mantle are from observations they make on the Earth's surface. In some places mantle rock has been pushed up to the surface by tectonic forces, allowing scientists to observe the rock directly.

Figure 2 *Volcanic vents on the ocean floor, such as this one off the coast of Hawaii, allow magma to escape from the mantle beneath oceanic crust.*

As you can see in **Figure 2,** another place scientists look is on the ocean floor, where molten rock from the mantle flows out of active volcanoes. These underwater volcanoes are like windows through the crust into the mantle. The "windows" have given us strong clues about the composition of the mantle. Scientists have learned that the mantle's composition is similar to that of the mineral olivine, which has large amounts of iron and magnesium compared with other common minerals.

The Core By studying the different layers that make up the Earth, geologists can get an idea of which elements each is made of. They think that the Earth's *core* is made mostly of iron, with smaller amounts of nickel and possibly some sulfur and oxygen. The **core** extends from the bottom of the mantle to the center of the Earth. As you can see in **Figure 3,** the diameter of the planet Mars is slightly smaller than that of the Earth's core.

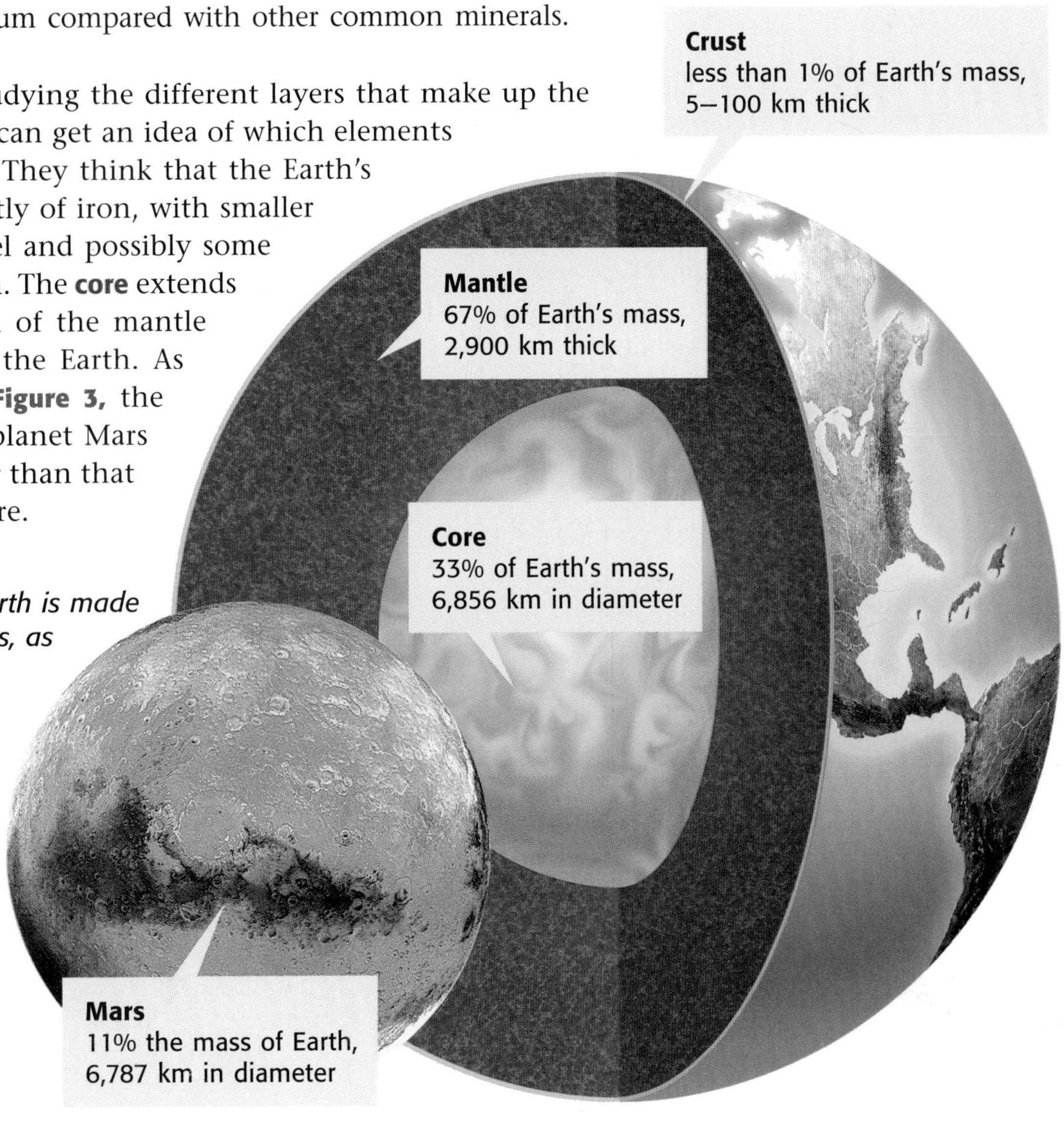

Crust
less than 1% of Earth's mass, 5–100 km thick

Mantle
67% of Earth's mass, 2,900 km thick

Core
33% of Earth's mass, 6,856 km in diameter

Mars
11% the mass of Earth, 6,787 km in diameter

Figure 3 *The Earth is made up of three layers, as shown here.*

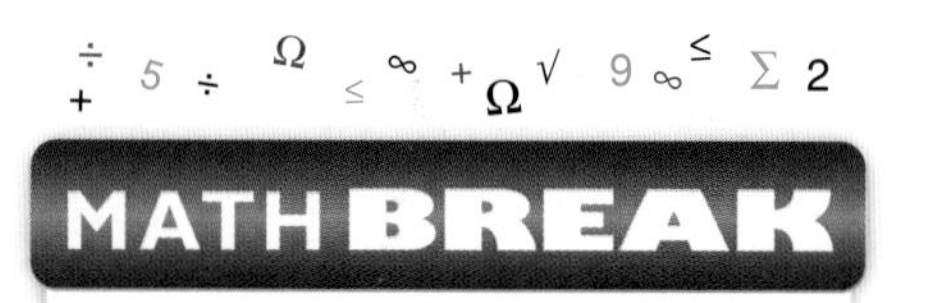

MATH**BREAK**

Using Models

Imagine that you are building a model of the Earth that is going to have a radius of 1 m. You find out that the average radius of the Earth is 6,378 km and that the thickness of the lithosphere is about 150 km. What percentage of the Earth's radius is the lithosphere? How thick (in centimeters) would you make the lithosphere in your model?

The Structure of the Earth

So far we have talked about the composition of the Earth. Another way to look at how the Earth is made is to examine the physical properties of its layers. The Earth is divided into five main physical layers—the *lithosphere, asthenosphere, mesosphere, outer core,* and *inner core.* As shown below, each layer has its own set of physical properties.

Lithosphere The outermost, rigid layer of the Earth is called the **lithosphere** ("rock sphere"). The lithosphere is made of two parts—the crust and the rigid upper part of the mantle. The lithosphere is divided into pieces called *tectonic plates.*

Asthenosphere The **asthenosphere** ("weak sphere") is a soft layer of the mantle on which pieces of the lithosphere move. It is made of solid rock that, like putty, flows very slowly—at about the same rate your fingernails grow.

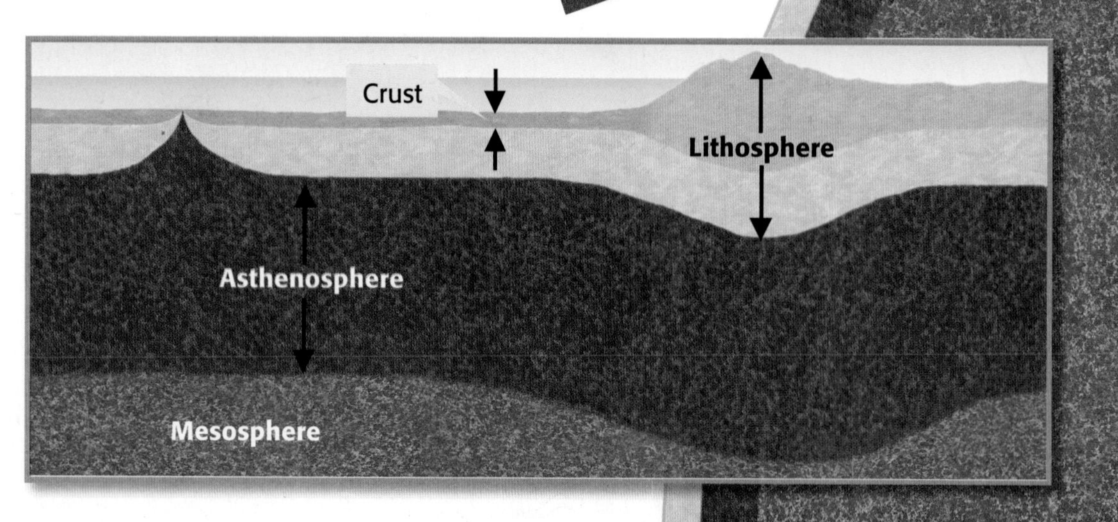

Scientists call the part of the Earth where life is possible the *biosphere.* The biosphere is the layer of the Earth above the crust and below the uppermost part of the atmosphere. It includes the oceans, the land surface, and the lower part of the atmosphere.

Mesosphere Beneath the asthenosphere is the strong, lower part of the mantle called the **mesosphere** ("middle sphere"). The mesosphere extends from the bottom of the asthenosphere down to the Earth's core.

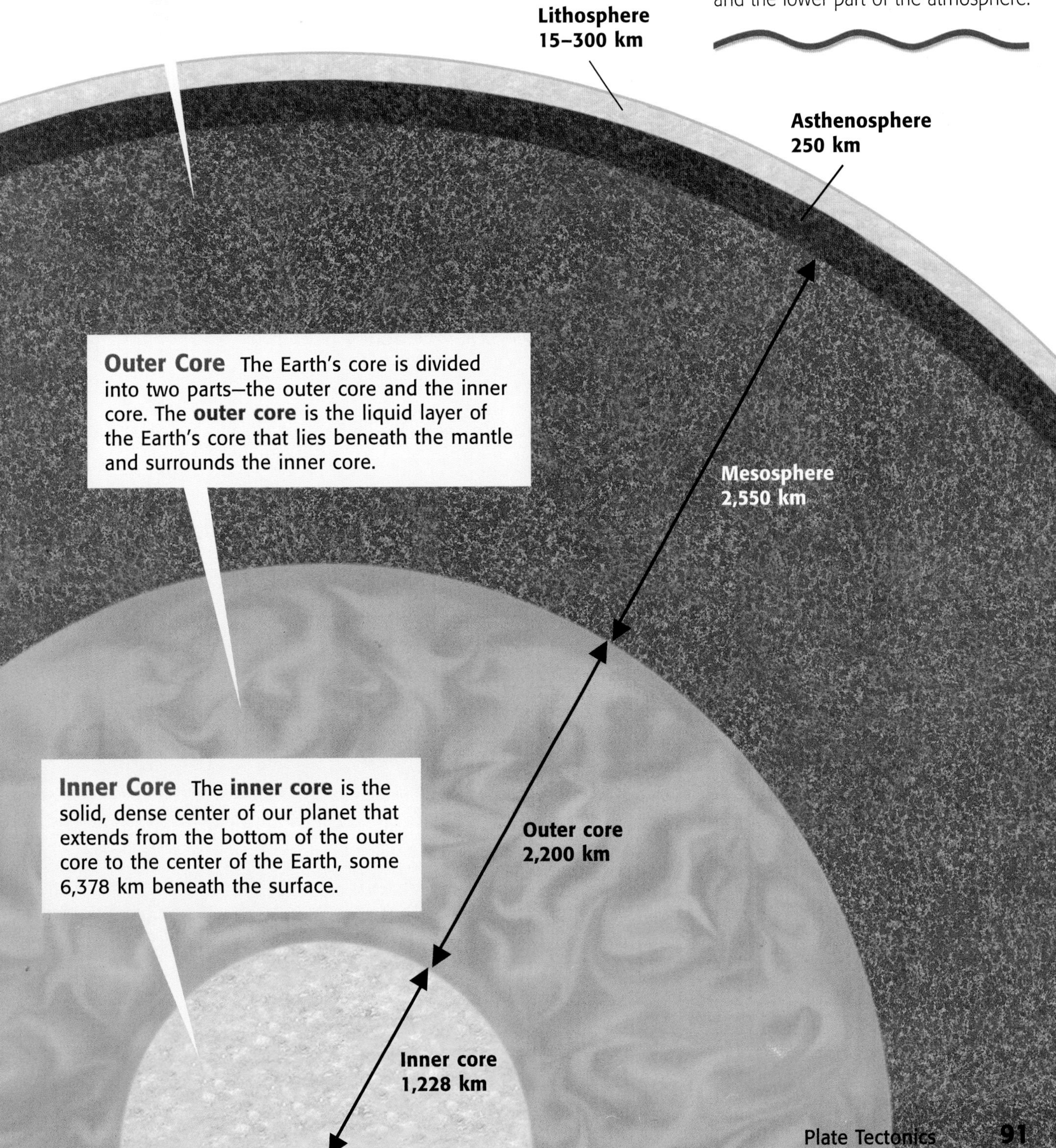

Outer Core The Earth's core is divided into two parts—the outer core and the inner core. The **outer core** is the liquid layer of the Earth's core that lies beneath the mantle and surrounds the inner core.

Inner Core The **inner core** is the solid, dense center of our planet that extends from the bottom of the outer core to the center of the Earth, some 6,378 km beneath the surface.

Tectonic Plates

Tectonic plates are pieces of the lithosphere that move around on top of the asthenosphere. But what exactly does a tectonic plate look like? How big are tectonic plates? How and why do they move around? To answer these questions, start by thinking of the lithosphere as a giant jigsaw puzzle.

Figure 4 *Tectonic plates fit together like the pieces of a jigsaw puzzle. On this map, the relative motions of some of the major tectonic plates are shown with arrows.*

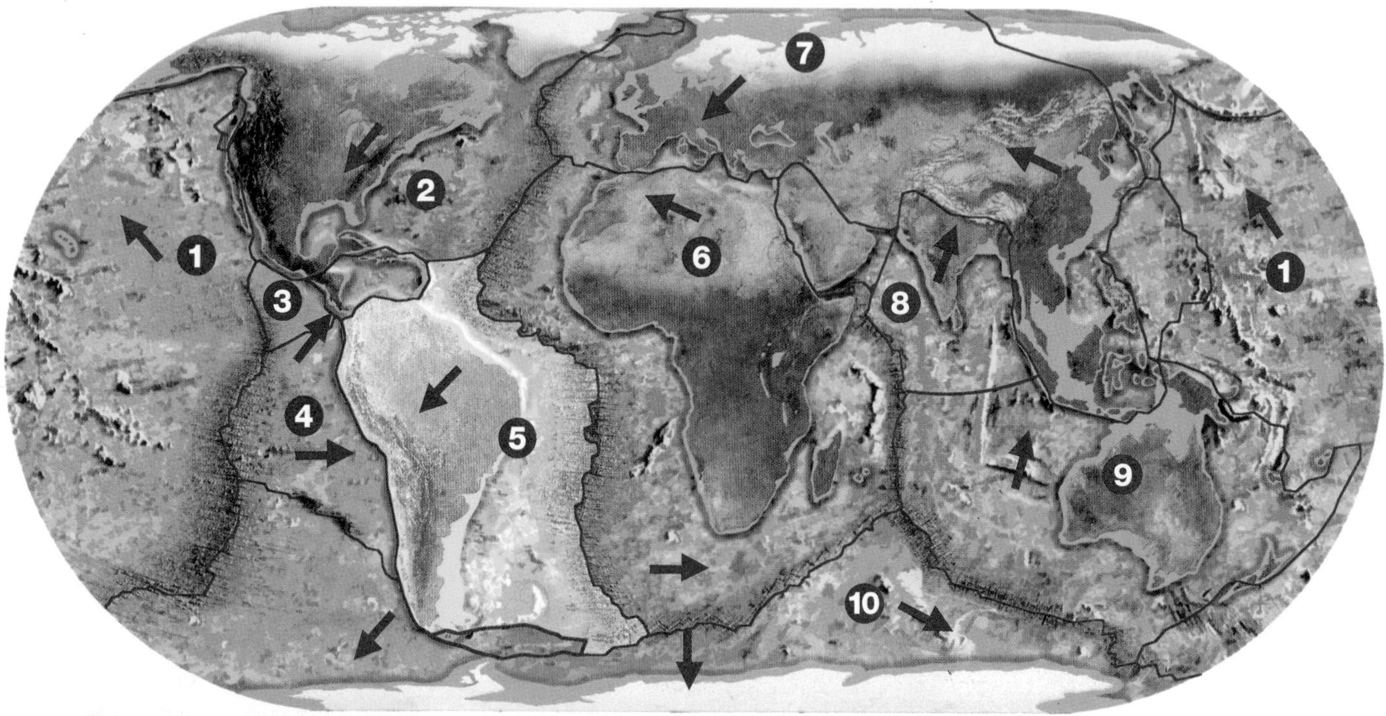

Major Tectonic Plates

1. Pacific plate
2. North American plate
3. Cocos plate
4. Nazca plate
5. South American plate
6. African plate
7. Eurasian plate
8. Indian plate
9. Australian plate
10. Antarctic plate

A Giant Jigsaw Puzzle Look at the world map above. All of the plates have names, some of which you may already be familiar with. Some of the major tectonic plates are listed in the key at left. Notice that each tectonic plate fits the other tectonic plates that surround it. The lithosphere is like a jigsaw puzzle, and the tectonic plates are like the pieces of a jigsaw puzzle.

You will also notice that not all tectonic plates are the same. Compare the size of the North American plate with that of the Cocos plate. But tectonic plates are different in other ways too. For example, the North American plate has an entire continent on it, while the Cocos plate only has oceanic crust. Like the North American plate, some tectonic plates include both continental *and* oceanic crust.

A Tectonic Plate Close-up What would a tectonic plate look like if you could lift it out of its place? **Figure 5** shows what the South American plate might look like if you could. Notice that this tectonic plate consists of both oceanic and continental crust, just like the North American plate.

The thickest part of this tectonic plate is on the South American continent, under the Andes mountain range. The thinnest part of the South American plate is at the Mid-Atlantic Ridge.

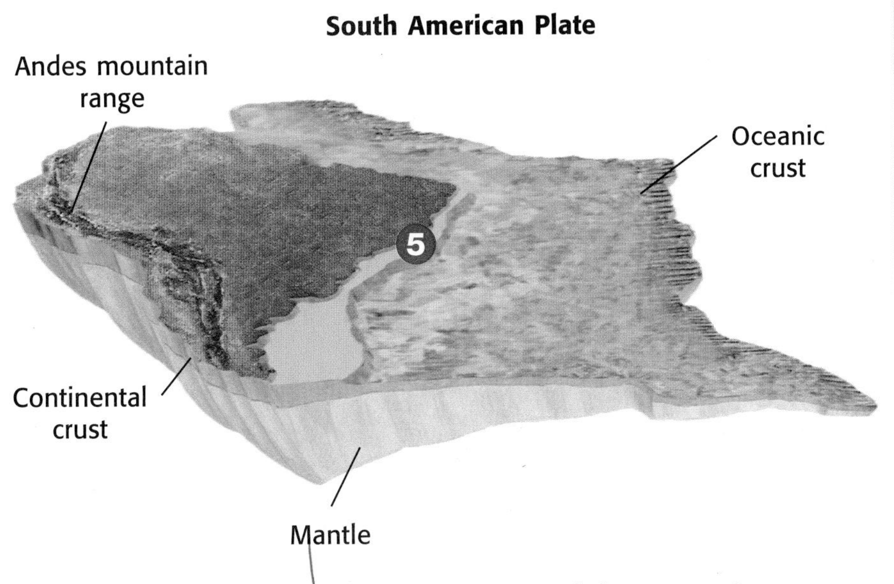

Figure 5 *The South American plate is one of the many pieces of the spherical "jigsaw puzzle" we call the lithosphere.*

Tip of the Iceberg If you could look at a tectonic plate from the side, you would see that mountain ranges are like the tips of icebergs—there is much more material below the surface than above. Mountain ranges that occur in continental crust have very deep roots relative to their height. For example, the Rocky Mountains rise less than 5 km above sea level, but their roots go down to about 60 km *below* sea level.

But if continental crust is so much thicker than oceanic crust, why doesn't it sink down below the oceanic crust? Think back to the difference between continental and oceanic crust. Continental crust stands much higher than oceanic crust because it is both thicker and less dense. Both kinds of crust are less dense than the mantle and "float" on top of the asthenosphere, similar to the way ice floats on top of water.

Floating Mountains

1. Take a large **block** of wood and place it in a clear plastic **container.** The block of wood represents the mantle part of the lithosphere.
2. Fill the container with **water** at least 10 cm deep. The water represents the asthenosphere. Use a ruler to measure how far the top of the wood block sits above the surface of the water.
3. Now try loading the block of wood with several different **wooden objects,** each with a different weight. These objects represent different amounts of crustal material loaded onto the lithosphere during mountain building. Measure how far the block sinks under each different weight.
4. What can you conclude about how the tectonic plate reacts to increasing weight of crustal material?
5. What happens to a tectonic plate when the crustal material is removed?

TRY at HOME

Mapping the Earth's Interior

How do we know all these things about the deepest parts of the Earth, where no one has ever been? Scientists have never even drilled through the crust, which is only a thin skin on the surface of the Earth. So how do we know so much about the mantle and the core?

Would you be surprised to know that the answers come from earthquakes? When an earthquake occurs, vibrations called seismic waves are produced. *Seismic waves* are vibrations that travel through the Earth. Depending on the density and strength of material they pass through, seismic waves travel at different speeds. For example, a seismic wave traveling through solid rock will go faster than a seismic wave traveling through a liquid.

When an earthquake occurs, *seismographs* measure the difference in the arrival times of seismic waves and record them. Seismologists can then use these measurements to calculate the density and thickness of each physical layer of the Earth. **Figure 6** shows how one kind of seismic wave travels through the Earth.

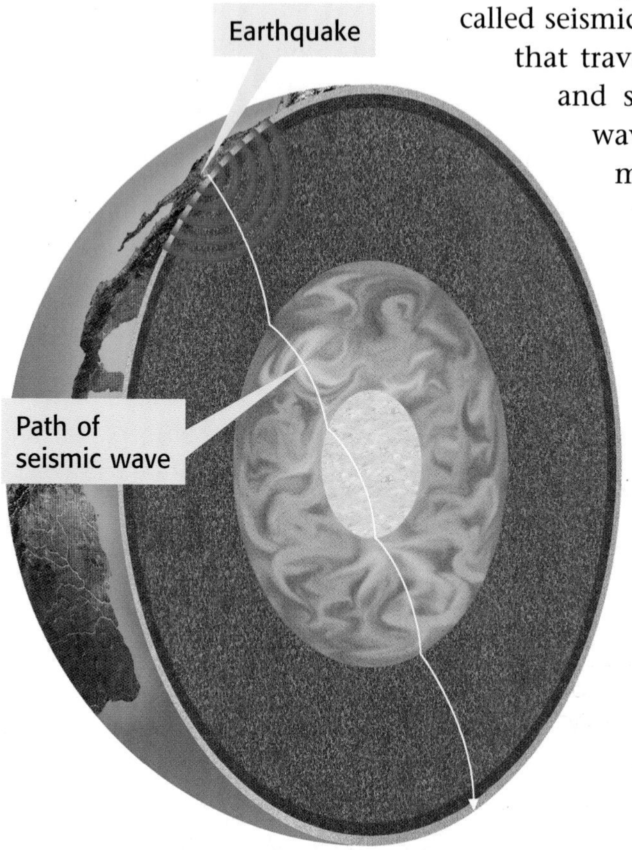

Lithosphere 7–8 km/second

Asthenosphere 7–11 km/second

Mesosphere 11–13 km/second

Outer core 7–10 km/second

Inner core 11–12 km/second

Figure 6 *The speed of seismic waves depends on the density of the material they travel through. The denser the material, the faster seismic waves move.*

internet**connect**

sciLINKS NSTA

TOPIC: Composition of the Earth, Structure of the Earth

GO TO: www.scilinks.org

***sci*LINKS NUMBER:** HSTE155, HSTE160

SECTION REVIEW

1. What is the difference between continental and oceanic crust?
2. How is the lithosphere different from the asthenosphere?
3. How do scientists know about the structure of the Earth's interior? Explain.
4. **Analyzing Relationships** Explain the difference between the crust and the lithosphere.

SECTION 2
READING WARM-UP

Restless Continents

Terms to Learn

continental drift
sea-floor spreading

What You'll Do

- Describe Wegener's theory of continental drift, and explain why it was not accepted at first.
- Explain how sea-floor spreading provides a way for continents to move.
- Describe how new oceanic crust forms at mid-ocean ridges.
- Explain how magnetic reversals provide evidence for sea-floor spreading.

Take a look at **Figure 7.** It shows how continents would fit together if you removed the Atlantic Ocean and moved the land together. Is it just coincidence that the coastlines fit together so well? Is it possible that the continents were actually together sometime in the past?

Africa
South America

Figure 7 *The theory of continental drift was inspired in part by the puzzlelike fit of the continents.*

Wegener's Theory of Continental Drift

One scientist who looked at the pieces of this puzzle was Alfred Wegener (VEG e nuhr). In the early 1900s he wrote about his theory of *continental drift.* **Continental drift** is the theory that continents can drift apart from one another and have done so in the past. This theory seemed to explain a lot of puzzling observations, including the very good fit of some of the continents.

Continental drift also explained why fossils of the same plant and animal species are found on both sides of the Atlantic Ocean. Many of these ancient species could not have made it across the Atlantic Ocean. As you can see in **Figure 8,** without continental drift, this pattern of fossil findings would be hard to explain. In addition to fossils, similar types of rock and evidence of the same ancient climatic conditions were found on several continents.

North America
Eurasia
India
Africa
South America
Australia
Antarctica
Mesosaurus
Glossopteris

Figure 8 *Fossils of* Mesosaurus, *a small, aquatic reptile, and* Glossopteris, *an ancient plant species, have been found on several continents.*

Continental drift also explained puzzling evidence left by ancient glaciers. Glaciers cut grooves in the ground that indicate the direction they traveled. When you look at the placement of today's continents, these glacial activities do not seem to be related. But when you bring all of these continental pieces back to their original arrangement, the glacial grooves match! Along with fossil evidence, glacial grooves supported Wegener's idea of continental drift.

The Breakup of Pangaea

Wegener studied many observations before establishing his theory of continental drift. He thought that all the separate continents of today were once joined in a single landmass that he called *Pangaea,* which is Greek for "all earth." As shown in **Figure 9,** almost all of Earth's landmasses were joined together in one huge continent 245 million years ago.

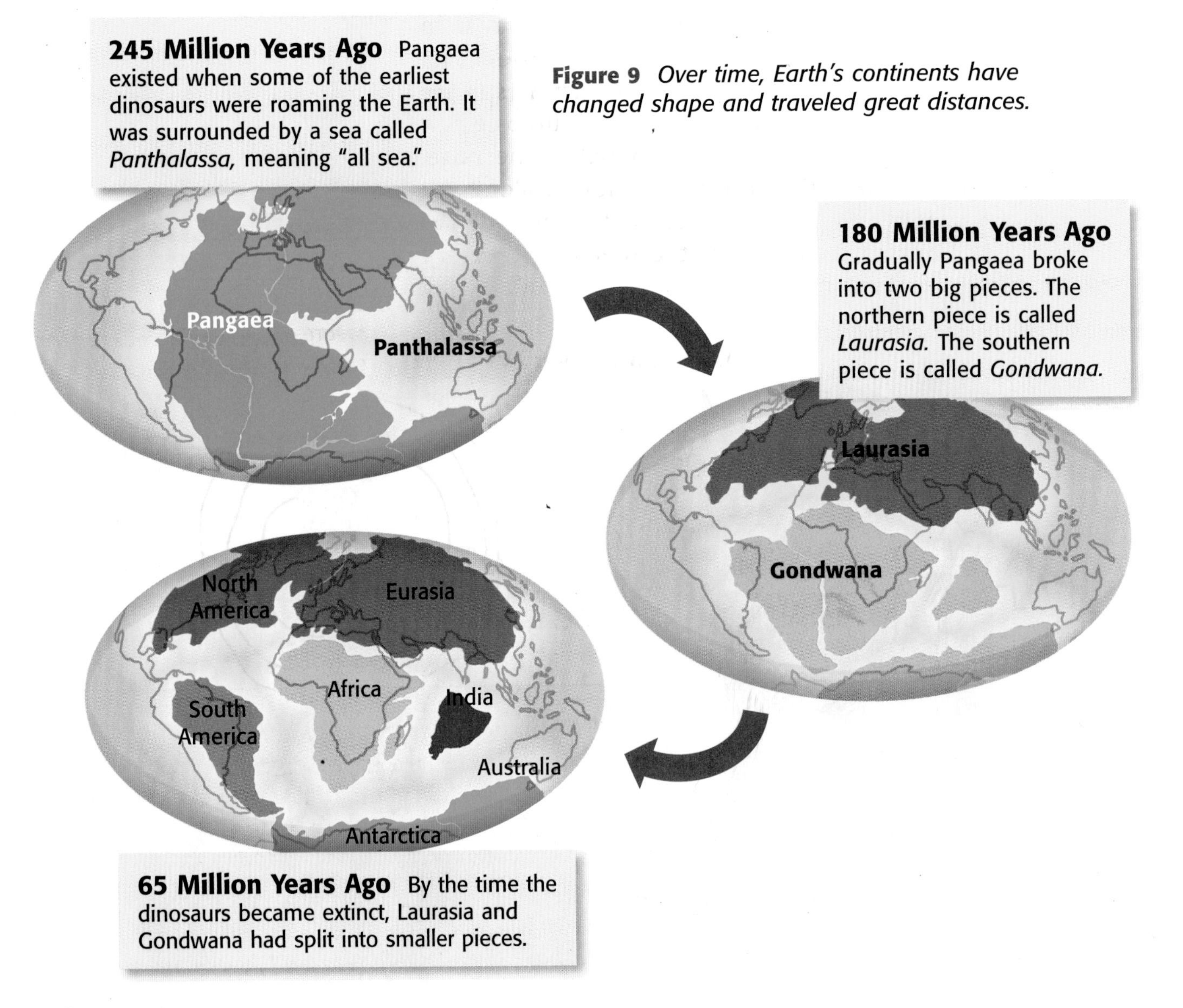

Figure 9 *Over time, Earth's continents have changed shape and traveled great distances.*

Sea-Floor Spreading

When Wegener put forth his theory of continental drift, many scientists would not accept his theory. What force of nature, they wondered, could move entire continents? In Wegener's day, no one could answer that question. It wasn't until many years later that new evidence provided some clues.

In **Figure 10** you will notice that there is a chain of submerged mountains running through the center of the Atlantic Ocean. The chain is called the Mid-Atlantic Ridge, part of a worldwide system of ocean ridges. Mid-ocean ridges are underwater mountain chains that run through Earth's ocean basins.

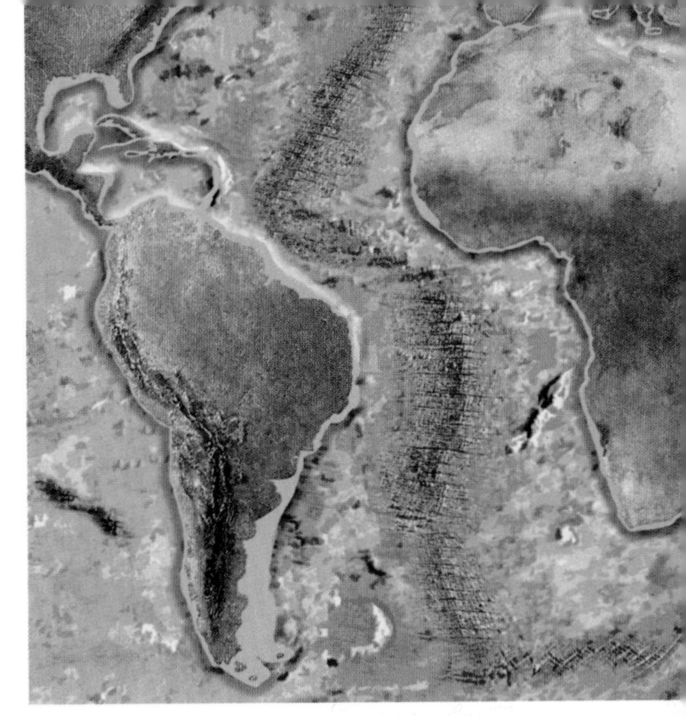

Figure 10 *The Mid-Atlantic Ridge is part of the longest mountain chain in the world.*

Mid-ocean ridges are places where sea-floor spreading takes place. **Sea-floor spreading** is the process by which new oceanic lithosphere is created as older materials are pulled away. As tectonic plates move away from each other, the sea floor spreads apart and magma rises to fill in the gap. Notice in **Figure 11** that the crust increases in age the farther it is from the mid-ocean ridge. This is because new crust continually forms from molten material at the ridge. The oldest crust in the Atlantic Ocean is found along the edges of the continents. It dates back to the time of the dinosaurs. The newest crust is in the center of the ocean. This crust has just formed!

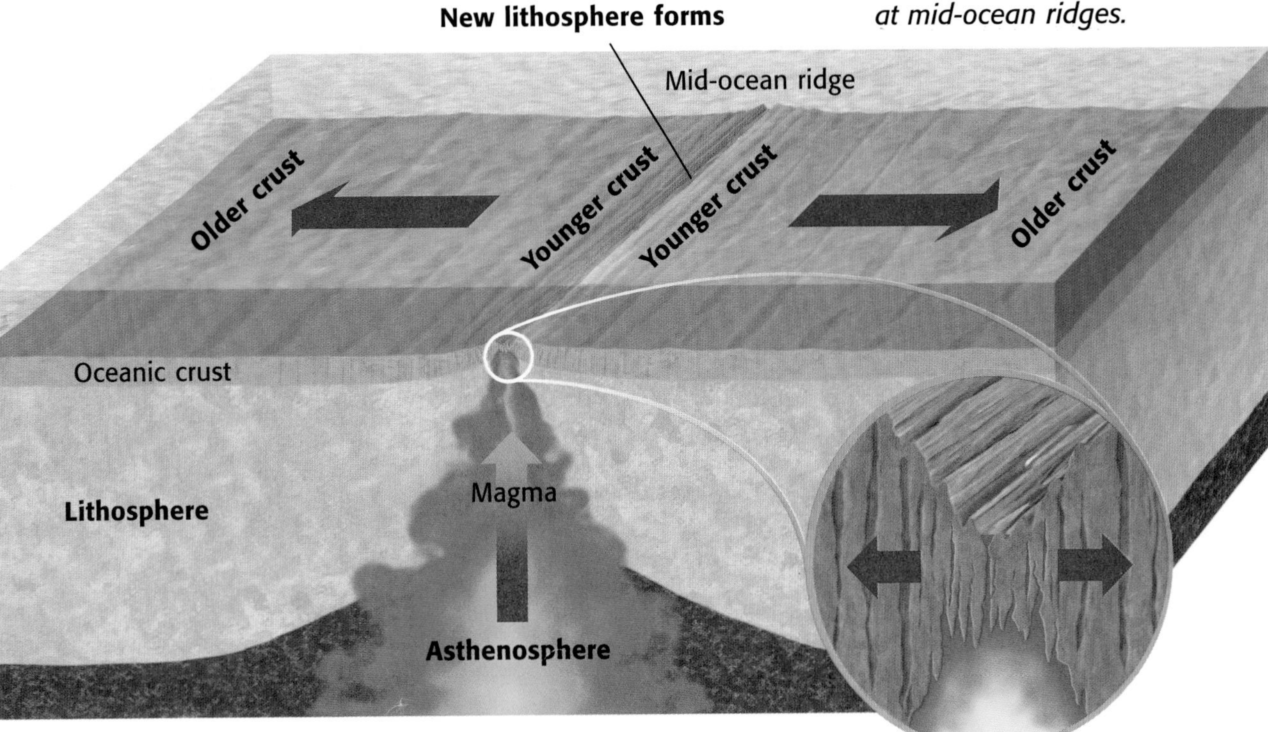

Figure 11 *Sea-floor spreading creates new oceanic lithosphere at mid-ocean ridges.*

All matter has the property of magnetism, though in most cases it is very weak compared with that of magnets. This explains why researchers have been able to levitate a frog—by creating a very strong magnetic field beneath it!

Magnetic Reversals

Some of the most important evidence of sea-floor spreading comes from magnetic reversals recorded in the ocean floor. Throughout Earth's history, the north and south magnetic poles have changed places many times. When Earth's magnetic poles change place, this is called a *magnetic reversal.*

The molten rock at the mid-ocean ridges contains tiny grains of magnetic minerals. These mineral grains act like compasses. They align with the magnetic field of the Earth. Once the molten rock cools, the record of these tiny compasses is literally set in stone. This record is then carried slowly away from the spreading center as sea-floor spreading occurs. As you can see in **Figure 12,** when the Earth's magnetic field reverses, a new band is started, and this time the magnetic mineral grains point in the opposite direction. The new rock records the direction of the Earth's magnetic field. This record of magnetic reversals was the final proof that sea-floor spreading does occur.

Figure 12 *Magnetic reversals in oceanic crust are shown here as bands of light and dark blue oceanic crust.*

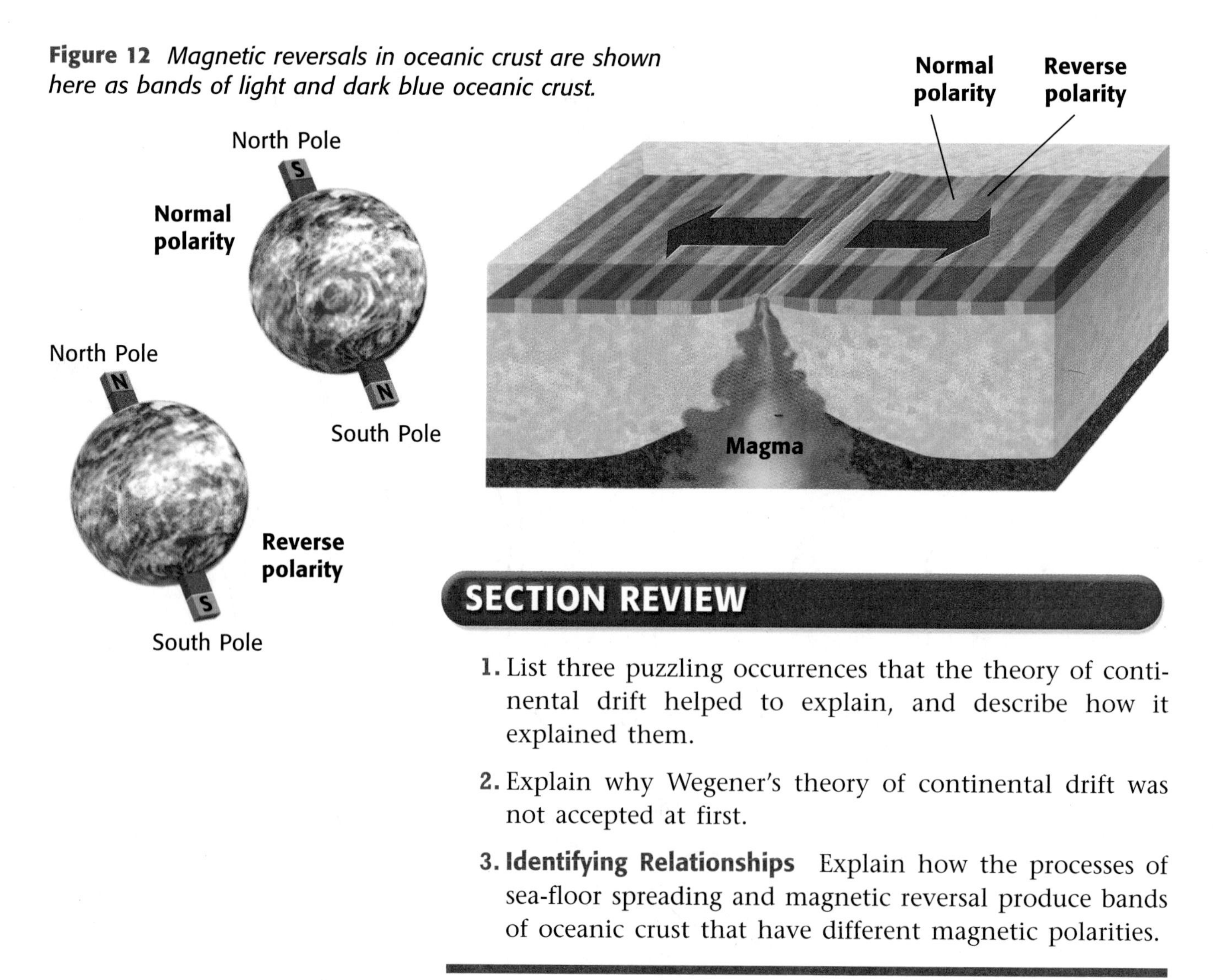

SECTION REVIEW

1. List three puzzling occurrences that the theory of continental drift helped to explain, and describe how it explained them.
2. Explain why Wegener's theory of continental drift was not accepted at first.
3. **Identifying Relationships** Explain how the processes of sea-floor spreading and magnetic reversal produce bands of oceanic crust that have different magnetic polarities.

SECTION 3

READING WARM-UP

Terms to Learn

plate tectonics
convergent boundary
subduction zone
divergent boundary
transform boundary

What You'll Do

- Describe the three forces thought to move tectonic plates.
- Describe the three types of tectonic plate boundaries.
- Explain how scientists measure the rate at which tectonic plates move.

The Theory of Plate Tectonics

The proof of sea-floor spreading supported Wegener's original idea that the continents move. But because both oceanic and continental crust appear to move, a new theory was devised to explain both continental drift and sea-floor spreading—the theory of *plate tectonics*. **Plate tectonics** is the theory that the Earth's lithosphere is divided into tectonic plates that move around on top of the asthenosphere.

Possible Causes of Tectonic Plate Motion

An incredible amount of energy is needed to move something as massive as a tectonic plate! We still don't know exactly why tectonic plates move as they do, but recently scientists have come up with some possible answers, as shown in **Figure 13.** Notice how all three are affected by heat and gravity.

Figure 13 Three Possible Driving Forces of Plate Tectonics

Ridge Push At mid-ocean ridges, the oceanic lithosphere is higher than it is where it sinks beneath continental lithosphere. *Ridge push* is the process by which an oceanic plate slides down the lithosphere-asthenosphere boundary.

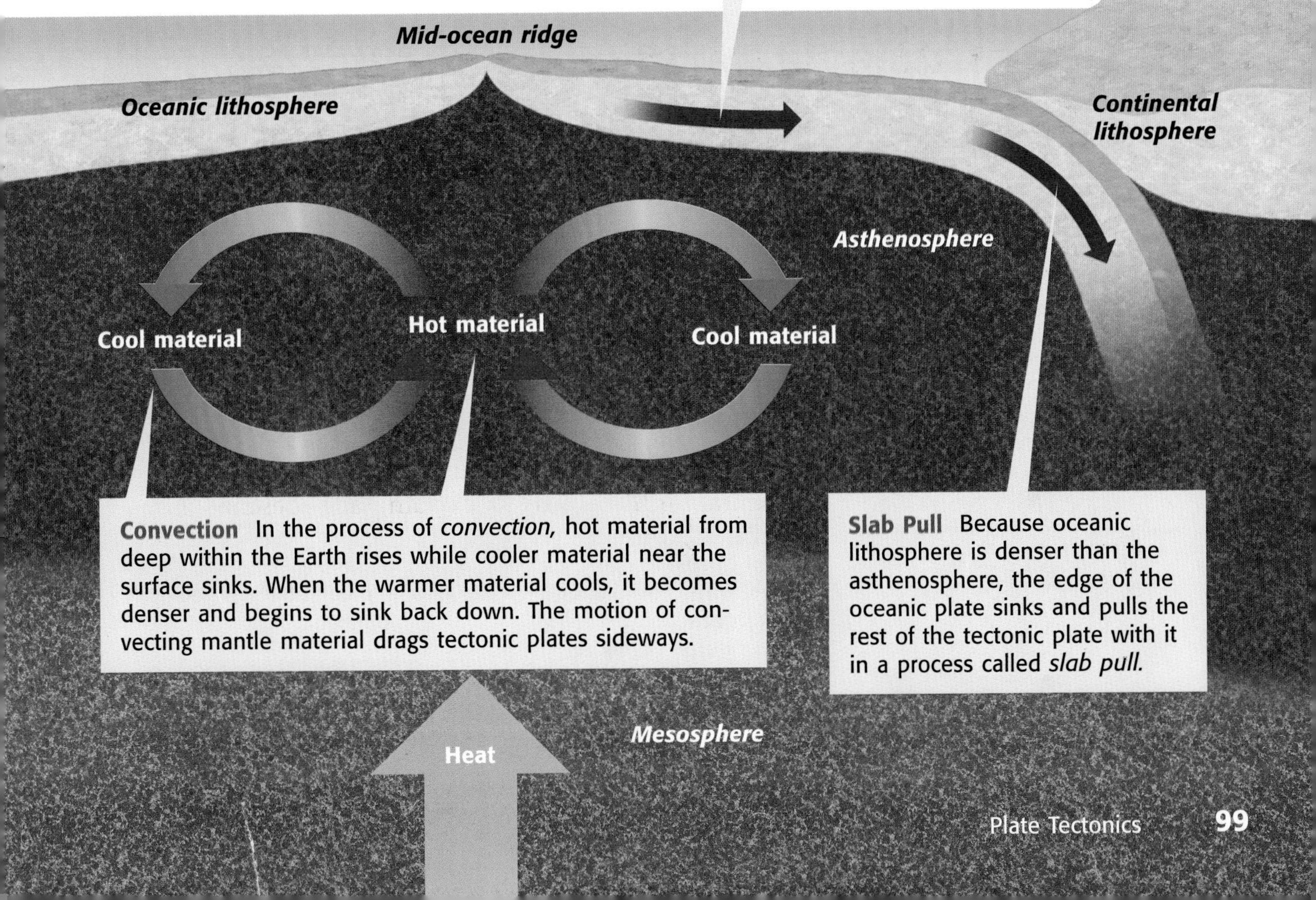

Mid-ocean ridge
Oceanic lithosphere
Continental lithosphere
Asthenosphere
Cool material
Hot material
Cool material

Convection In the process of *convection,* hot material from deep within the Earth rises while cooler material near the surface sinks. When the warmer material cools, it becomes denser and begins to sink back down. The motion of convecting mantle material drags tectonic plates sideways.

Slab Pull Because oceanic lithosphere is denser than the asthenosphere, the edge of the oceanic plate sinks and pulls the rest of the tectonic plate with it in a process called *slab pull.*

Mesosphere
Heat

Tectonic Plate Boundaries

All tectonic plates have boundaries with other tectonic plates. These boundaries are divided into three main types depending on how the tectonic plates move relative to one another. Tectonic plates can collide, separate, or slide past each other. **Figure 14** shows some examples of tectonic plate boundaries.

Convergent Boundaries When two tectonic plates push into one another, the boundary where they meet is called a **convergent boundary.** What happens at a convergent boundary depends on what kind of crust—continental or oceanic—the leading edge of each tectonic plate has. As you can see below, there are three types of convergent boundaries—continental/continental, continental/oceanic, and oceanic/oceanic.

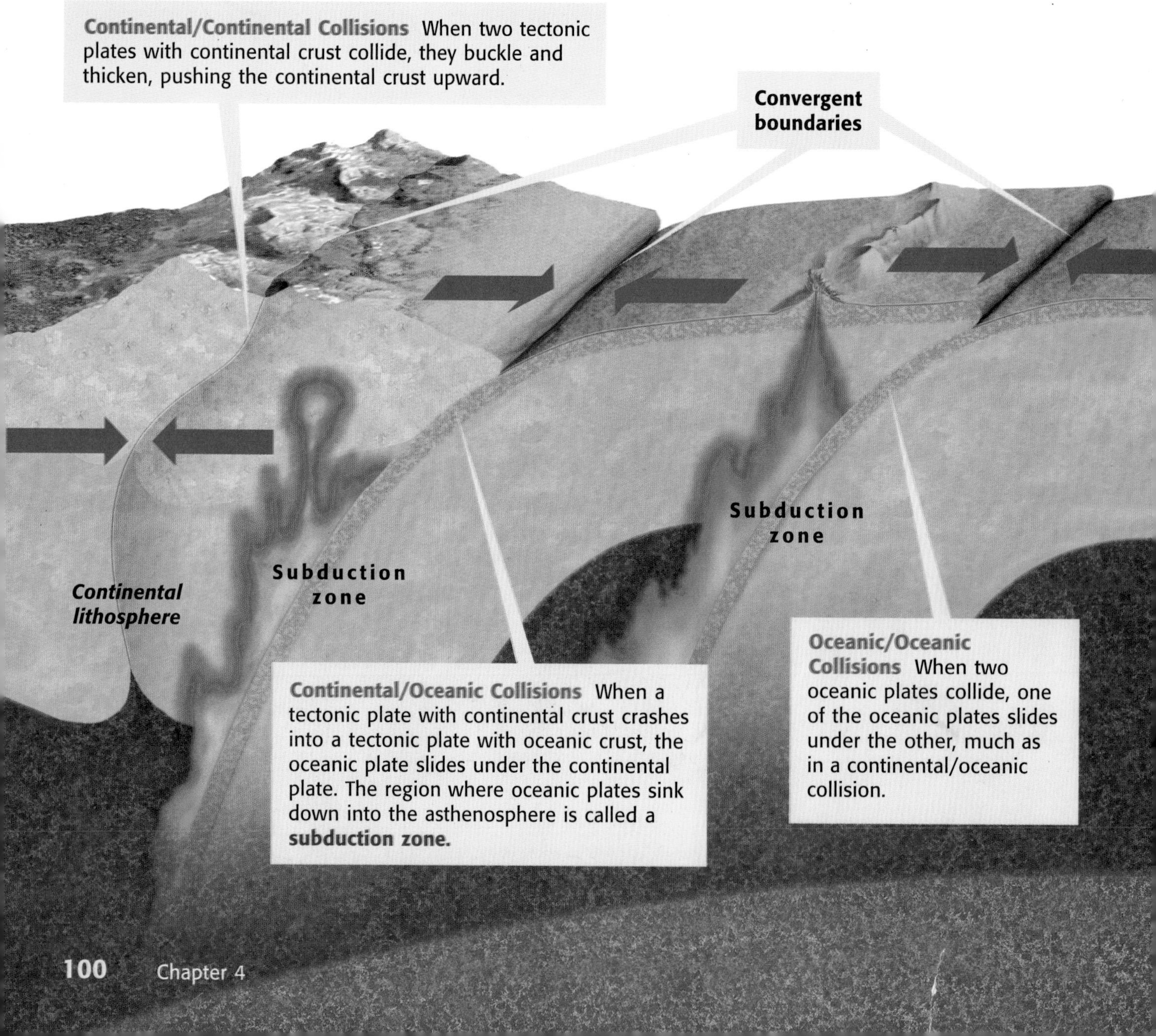

Figure 14 *This diagram shows five tectonic plate boundaries. Notice that there are three types of convergent boundaries.*

Continental/Continental Collisions When two tectonic plates with continental crust collide, they buckle and thicken, pushing the continental crust upward.

Continental/Oceanic Collisions When a tectonic plate with continental crust crashes into a tectonic plate with oceanic crust, the oceanic plate slides under the continental plate. The region where oceanic plates sink down into the asthenosphere is called a **subduction zone.**

Oceanic/Oceanic Collisions When two oceanic plates collide, one of the oceanic plates slides under the other, much as in a continental/oceanic collision.

Divergent Boundaries When two tectonic plates move away from one another, the boundary between them is called a **divergent boundary.** Remember sea-floor spreading? Divergent boundaries are where new oceanic lithosphere forms. The mid-ocean ridges that mark the spreading centers are the most common type of divergent boundary. However, divergent boundaries can also be found on continents.

Transform Boundaries When two tectonic plates slide past each other horizontally, the boundary between them is called a **transform boundary.** The San Andreas Fault, in California, is a good example of a transform boundary. This fault marks the place where the Pacific plate and the North American plate slide past each other.

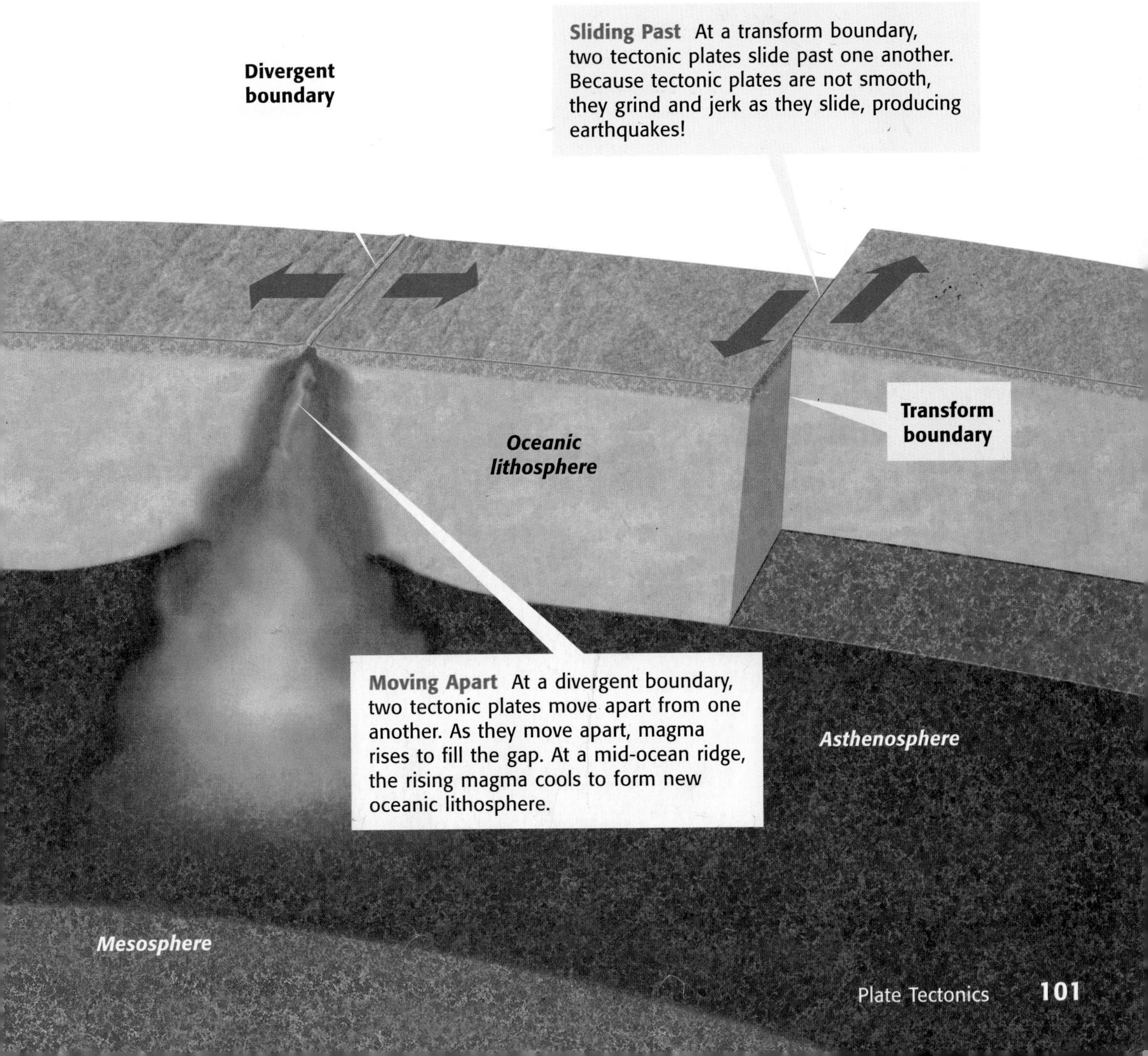

Sliding Past At a transform boundary, two tectonic plates slide past one another. Because tectonic plates are not smooth, they grind and jerk as they slide, producing earthquakes!

Divergent boundary

Transform boundary

Oceanic lithosphere

Moving Apart At a divergent boundary, two tectonic plates move apart from one another. As they move apart, magma rises to fill the gap. At a mid-ocean ridge, the rising magma cools to form new oceanic lithosphere.

Asthenosphere

Mesosphere

Tracking Tectonic Plate Motion

Just how fast do tectonic plates move? The answer to this question depends on many factors, such as the type of tectonic plate, the shape of the tectonic plate, and the way it interacts with the tectonic plates that surround it. Tectonic movements are generally so slow and gradual that you can't see or feel them—they are measured in centimeters per year.

One exception to this rule is the San Andreas Fault, in California. The Pacific plate and the North American plate do not slide past each other smoothly nor continuously. Instead, this movement happens in jerks and jolts. Sections of the fault remain stationary for years and then suddenly shift several meters, causing an earthquake. Large shifts that occur at the San Andreas fault can be measured right on the surface. Unfortunately for scientists, however, most movements of tectonic plates are very difficult to measure. So how do they do it?

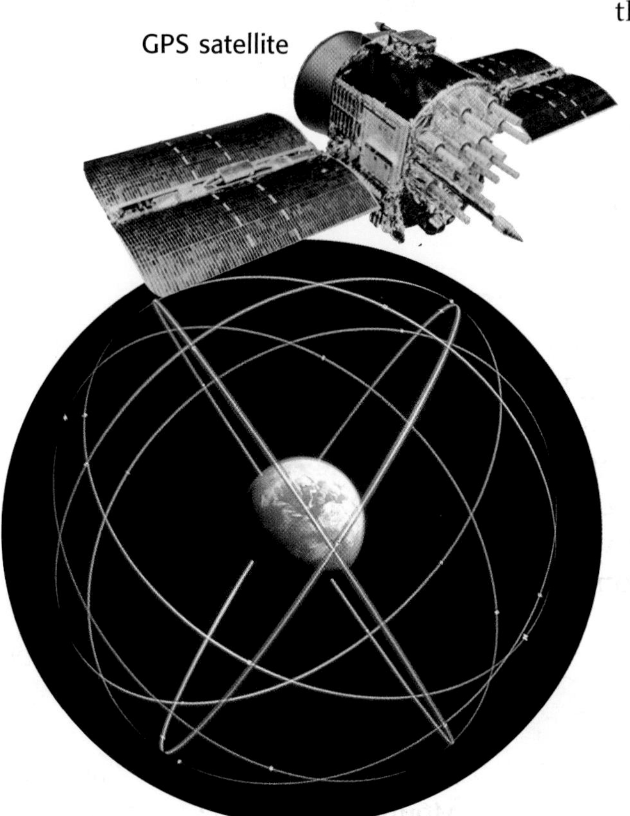

Figure 15 *The image above shows the orbits of the GPS satellites.*

The Global Positioning System Scientists use a network of satellites called the *Global Positioning System* (GPS), shown in **Figure 15,** to measure the rate of tectonic plate movement. Radio signals are continuously beamed from satellites to GPS ground stations, which record the exact distance between the satellites and the ground station. Over time, these distances change slightly. By recording the time it takes for the GPS ground stations to move a given distance, scientists can measure the rate of motion of each tectonic plate.

internet**connect**

TOPIC: Tectonic Plates
GO TO: www.scilinks.org
***sci*LINKS NUMBER:** HSTE165

SECTION REVIEW

1. List and describe three possible driving forces of tectonic plate motion.
2. How do the three types of convergent boundaries differ from one another?
3. Explain how scientists measure the rate at which tectonic plates move.
4. **Identifying Relationships** When convection takes place in the mantle, why does cooler material sink, while warmer material rises?

SECTION 4

READING WARM-UP

Deforming the Earth's Crust

Terms to Learn

stress
compression
tension
folding
fault
normal fault
reverse fault
strike-slip fault

What You'll Do

- Describe major types of folds.
- Explain how the three major types of faults differ.
- Name and describe the most common types of mountains.
- Explain how various types of mountains form.

Have you ever tried to bend something, only to have it break? Try this: take a long, uncooked piece of spaghetti, and bend it very slowly, and only a little. Now bend it again, but this time much farther and faster. What happened to it the second time? How can the same material bend at one time and break at another? The answer is that the *stress* you put on it was different. **Stress** is the amount of force per unit area that is put on a given material. The same principle works on the rocks in the Earth's crust. The conditions under which a rock is stressed determine its behavior.

Rocks Get Stressed

When rock changes its shape due to stress, this reaction is called *deformation.* In the example above, you saw the spaghetti deform in two different ways—by bending and by breaking. **Figure 16** illustrates this concept. The same thing happens in rock layers. Rock layers can bend when stress is placed on them. But when more stress is placed on them, they can break. Rocks can deform due to the forces of plate tectonics.

The type of stress that occurs when an object is squeezed, as when two tectonic plates collide, is called **compression.** Compression can have some spectacular results. The Rocky Mountains and the Cascade Range are two examples of compression at a convergent plate boundary.

Another form of stress is *tension.* **Tension** is stress that occurs when forces act to stretch an object. As you might guess, tension occurs at divergent plate boundaries, when two tectonic plates pull away from each other. In the following pages you will learn how these two tectonic forces—compression and tension—bend and break rock to form some of the common landforms you already know.

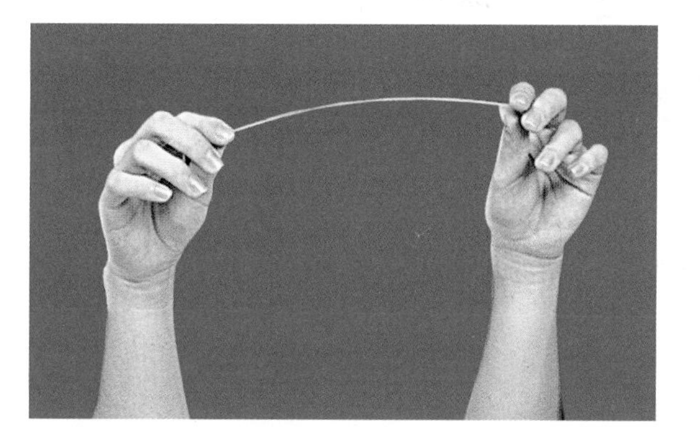

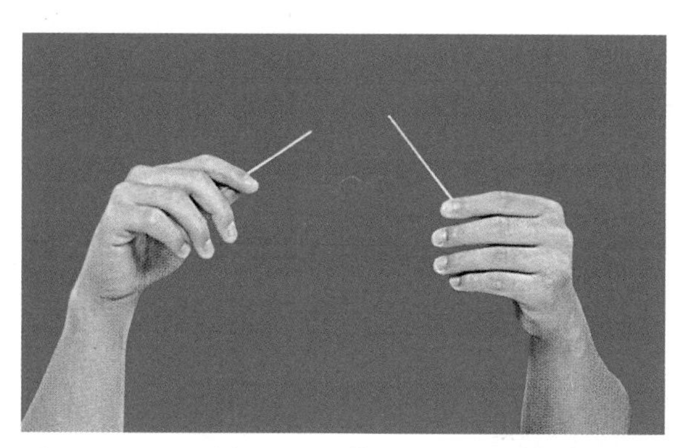

Figure 16 *With a small amount of stress, uncooked spaghetti bends. Additional stress causes it to break.*

Folding

Folding occurs when rock layers bend due to stress in the Earth's crust. We assume that all sedimentary rock layers started out as horizontal layers. So when you see a fold, you know that deformation has taken place. Depending on how the rock layers deform, different types of folds are made. **Figure 17** shows the two most common types—*anticlines* and *synclines*.

Another type of fold is a *monocline*. In a monocline, rock layers are folded so that both ends of the fold are still horizontal. Imagine taking a stack of paper and laying it on a table top. Think of all the sheets of paper as different rock layers. Now put a book under one end of the stack. You can see that both ends of the sheets are still horizontal, but all the sheets are bent in the middle.

Folds can be large or small. Take a look at **Figure 18.** The largest folds are measured in kilometers. They can make up the entire side of a mountain. Other folds are still obvious but much smaller. Note the size of the pocket knife in the smaller photo. Now look at the smallest folds. You would measure these folds in centimeters.

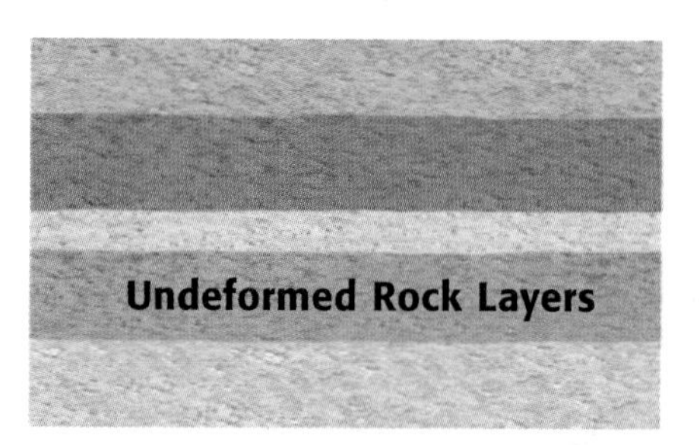

Figure 17 *When tectonic forces put stress on rock layers, they can cause the layers to bend and fold.* Anticlines *and* synclines *form when horizontal stress acts on rock.* Monoclines *form when vertical stress acts on rock.*

Figure 18 *The larger photo at right shows mountain-sized folds in the Rocky Mountains. The smaller photo shows a rock with much smaller folds.*

Faulting

While some rock layers bend and fold when stress is applied, other rock layers break. The surface along which rocks break and slide past each other is called a **fault.** The blocks of crust on each side of the fault are called *fault blocks.*

If a fault is not vertical, it is useful to distinguish between its two sides—the *hanging wall* and the *footwall.* **Figure 19** shows the difference between a hanging wall and a footwall. Depending on how the hanging wall and footwall move relative to each other, one of two main types of faults can form.

Normal Faults A *normal fault* is shown in **Figure 20.** The movement of a **normal fault** causes the hanging wall to move down relative to the footwall. Normal faults usually occur when tectonic forces cause tension that pulls rocks apart.

Reverse Faults A *reverse fault* is shown in **Figure 21.** The movement of a **reverse fault** causes the hanging wall to move up relative to the footwall—the "reverse" of a normal fault. Reverse faults usually happen when tectonic forces cause compression that pushes rocks together.

Self-Check

How is folding different from faulting? *(See page 216 to check your answer.)*

Figure 19 *The position of a fault block determines whether it is a hanging wall or a footwall.*

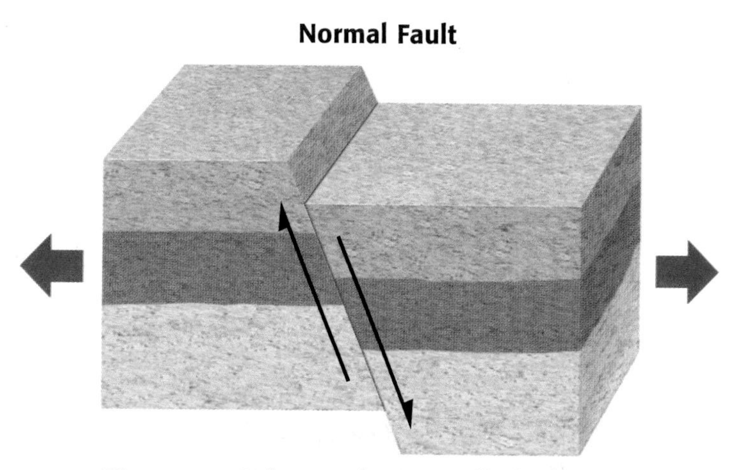

Figure 20 *When rocks are pulled apart due to tension, normal faults often result.*

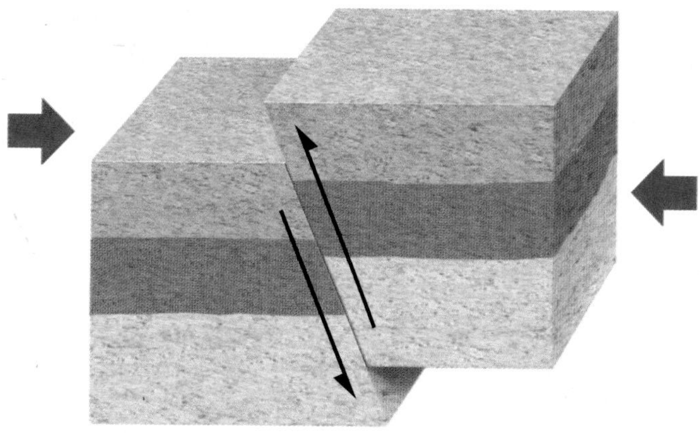

Figure 21 *When rocks are pushed together by compression, reverse faults often result.*

Figure 22 *The photo at left is a normal fault. The photo at right is a reverse fault.*

Telling the Difference It's easy to tell the difference between a normal fault and a reverse fault in diagrams with arrows. But what about the faults in **Figure 22?** You can certainly see the faults, but which one is a normal fault, and which one is a reverse fault? In the top left photo, one side has obviously moved relative to the other. You can tell this is a normal fault by looking at the sequence of sedimentary rock layers. You can see by the relative positions of the two dark layers that the hanging wall has moved down relative to the footwall.

Strike-slip Faults A third major type of fault is called a *strike-slip fault.* **Strike-slip faults** occur when opposing forces cause rock to break and move horizontally. If you were standing on one side of a strike-slip fault looking across the fault when it moved, the ground on the other side would appear to move to your left or right.

Tectonics and Natural Gas

Natural gas is used in many homes and factories as a source of energy. Some companies explore for sources of natural gas just as other companies explore for oil and coal. Like oil, natural gas travels upward through rock layers until it hits a layer through which it cannot travel and becomes trapped. Imagine that you are searching for pockets of trapped natural gas. Would you expect to find these pockets associated with anticlines, synclines, or faults? Explain your answer in your ScienceLog. Include drawings to help in your explanation.

Plate Tectonics and Mountain Building

You have just learned about several ways the Earth's crust changes due to the forces of plate tectonics. When tectonic plates collide, land features that start out as small folds and faults can eventually become great mountain ranges. The reason mountains exist is that tectonic plates are continually moving around and bumping into one another. As you can see in **Figure 23,** most major mountain ranges form at the edges of tectonic plates.

When tectonic plates undergo compression or tension, they can form mountains in several different ways. Let's take a look at three of the most common types of mountains—*folded mountains, fault-block mountains,* and *volcanic mountains.*

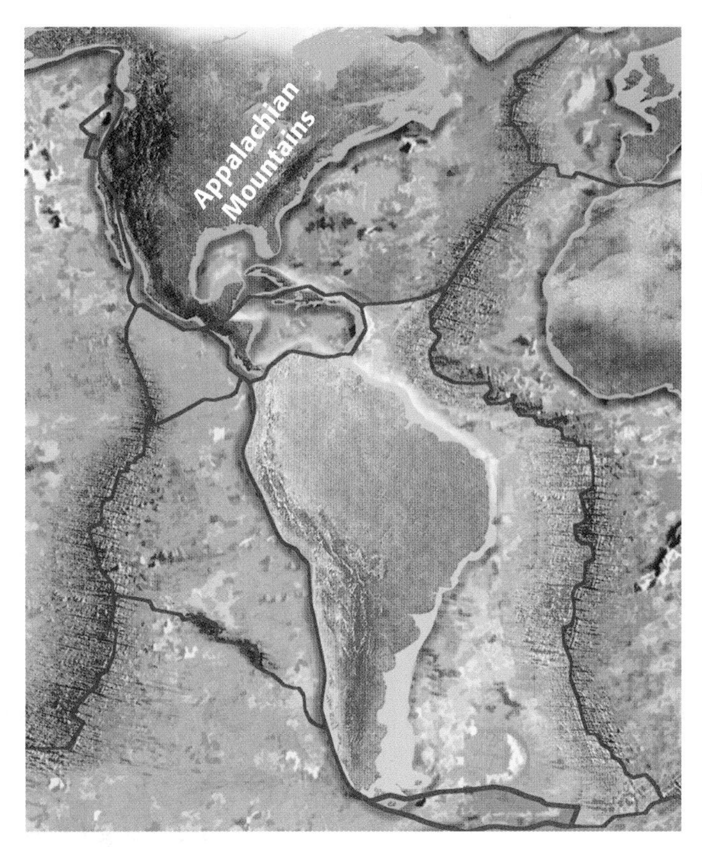

Figure 23 *Most of the world's major mountain ranges form at tectonic plate boundaries. Notice that the Appalachian Mountains, however, are located in the middle of the North American plate.*

Folded Mountains *Folded mountains* form when rock layers are squeezed together and pushed upward. If you take a pile of paper on a table top and push on opposite edges of the pile, you will see how a folded mountain forms. You saw how these layers crunched together in Figure 17. **Figure 24** shows an example of a folded mountain range that formed at a convergent boundary.

Figure 24 *Once as mighty as the Himalayas, the Appalachians have been worn down by hundreds of millions of years of weathering and erosion.*

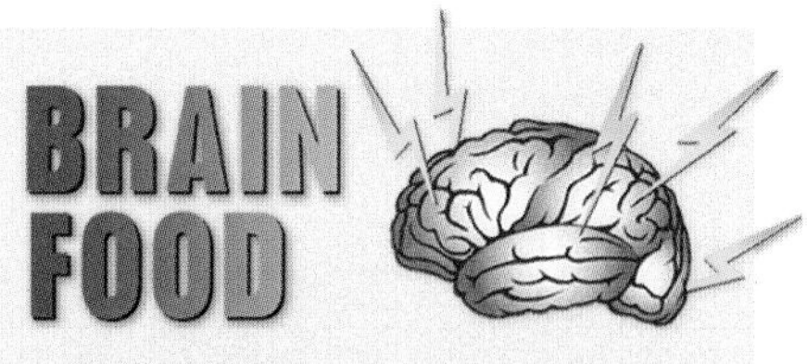

Did you know that plate tectonics is responsible for creating not only mountains but some of the lowest places on Earth as well? It's true. When one tectonic plate is subducted beneath another, a deep valley called a *trench* forms at the boundary. The Mariana Trench is the deepest point in the oceans—11,033 m below sea level!

Formation of the Appalachian Mountains

Look back at Figure 23. The Appalachians are in the middle of the North American plate. How can this be? Shouldn't they be at the edge of a tectonic plate? Follow along in this diagram to find the answer.

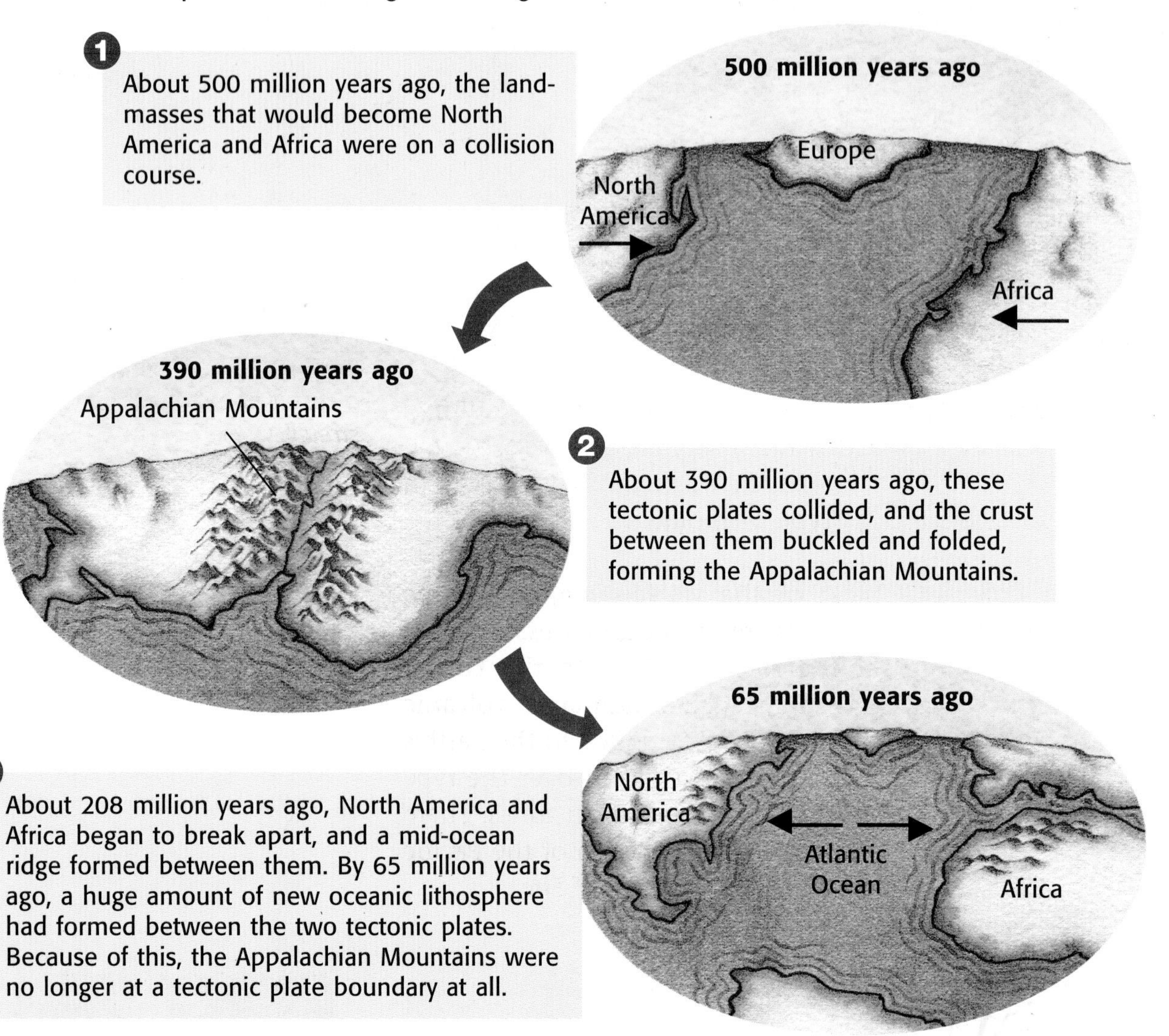

Figure 25 *When the crust is subjected to tension, the rock can break along a series of normal faults, resulting in fault-block mountains.*

Fault-block Mountains Where tectonic forces put enough tension on the Earth's crust, a large number of normal faults can result. *Fault-block mountains* form when this faulting causes large blocks of the Earth's crust to drop down relative to other blocks. **Figure 25** shows one way this can happen.

Figure 26 *The Tetons formed as a result of tectonic forces that stretched the Earth's crust, causing it to break in a series of normal faults. Compare this photo with the illustration in Figure 25.*

When sedimentary rock layers are tilted up by faulting, they can produce mountains with sharp, jagged peaks. As you can see in **Figure 26,** the Tetons, in western Wyoming, are a spectacular example of this type of mountain.

Volcanic Mountains Most of the world's major volcanic mountains are located at convergent boundaries. *Volcanic mountains* form when molten rock erupts onto the Earth's surface. Unlike folded and fault-block mountains, volcanic mountains form from new material being added to the Earth's surface. Most volcanic mountains tend to form over the type of convergent boundaries that include subduction zones. There are so many volcanic mountains around the rim of the Pacific Ocean that early explorers named it the *Ring of Fire.*

SECTION REVIEW

1. What is the difference between an anticline and a syncline?
2. What is the difference between a normal fault and a reverse fault?
3. Name and describe the type of tectonic stress that forms folded mountains.
4. Name and describe the type of tectonic stress that forms fault-block mountains.
5. **Making Predictions** If a fault occurs in an area where rock layers have been folded, which type of fault is it likely to be? Why?

Making Models Lab

USING SCIENTIFIC METHODS

Oh, the Pressure!

When scientists want to understand natural processes, such as mountain formation, they often make models to help them. Models are useful in studying how rocks react to the forces of plate tectonics. In a short amount of time, a model can demonstrate geological processes that take millions of years. Do the following activity to find out how folding and faulting happen in the Earth's crust.

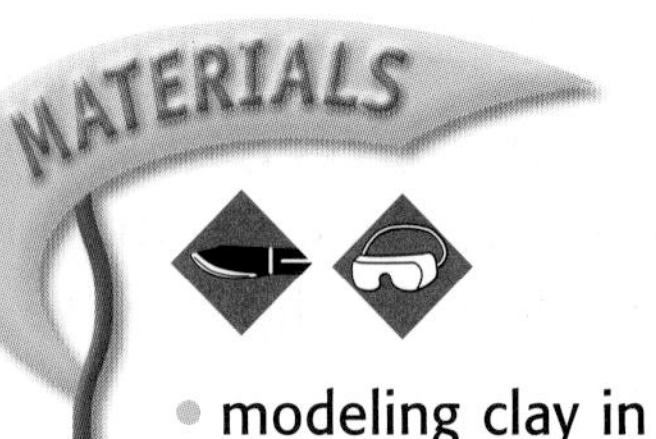

- modeling clay in 4 different colors
- 5 cm × 15 cm strip of poster board
- soup can or rolling pin
- newspaper
- colored pencils
- plastic knife
- 5 cm × 5 cm squares of poster board (2)

Ask a Question

1. What types of forces cause synclines, anticlines, and faults to form?

Make a Model

2. Use modeling clay of one color to form a long cylinder. Place the cylinder in the center of the glossy side of the poster-board strip.
3. Mold the clay to the strip. Try to make the clay layer the same thickness all along the strip; you can use the soup can or rolling pin to even the clay out. Pinch the sides of the clay so that it is the same width and length as the strip. Your strip should be at least 15 cm long and 5 cm wide.

4. Flip the strip over on the newspaper your teacher has placed across your desk. Carefully peel the strip from the modeling clay.

5. Repeat steps 2–4 with the other colors of modeling clay. Each member of your group should have a turn molding the clay. Each time you flip the strip over, stack the new clay layer on top of the previous one. When you are finished, you should have a block of clay made of four layers.

6. Lift the block of clay and hold it parallel to and just above the table top. Push gently on the block from opposite sides, as shown on the previous page.

7. Use the colored pencils to draw the results of step 6 in your ScienceLog. Use the terms *syncline* and *anticline* to label your diagram. Draw arrows to show the direction that each edge of the clay was pushed.

8. Repeat steps 2–5 to form a second block of clay.

9. Using the plastic knife, carefully cut the second block of clay in two at a 45° angle as seen from the side of the block.

10. Press one poster-board square on the angled end of each of the block's two pieces. The poster board represents a fault. The two angled ends represent a hanging wall and a footwall. The model should resemble the one in the photograph below.

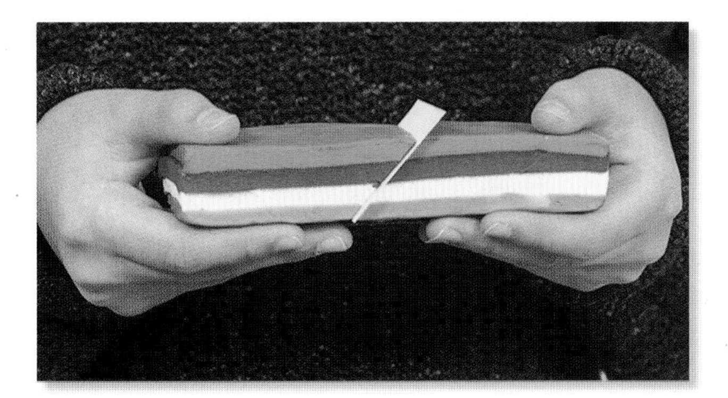

11. Keeping the angled edges together, lift the blocks and hold them parallel to and just above the table top. Push gently on the two blocks until they move. Record your observations in your ScienceLog.

12. Now hold the two pieces of the clay block in their original position, and slowly pull them apart, allowing the two blocks to slide past each other. Record your observations.

Analyze the Results

13. What happened to the first block of clay in step 6? What force did you model?

14. What happened to the pieces of the second block of clay in step 11? What force did you model?

15. What happened to the pieces of the second block of clay in step 12? Describe the forces that acted on the block and how the pieces of the block reacted.

Draw Conclusions

16. Summarize how the forces you applied to the blocks of clay relate to the way tectonic forces affect rock layers. In communicating your results, be sure to use the terms *fold, fault, anticline, syncline, hanging wall, footwall, tension,* and *compression.*

Chapter Highlights

SECTION 1

Vocabulary

crust *(p. 88)*
mantle *(p. 89)*
core *(p. 89)*
lithosphere *(p. 90)*
asthenosphere *(p. 90)*
mesosphere *(p. 91)*
outer core *(p. 91)*
inner core *(p. 91)*
tectonic plate *(p. 92)*

Section Notes

- The Earth is made of three basic compositional layers—the crust, the mantle, and the core.
- The Earth is made of five main structural layers—lithosphere, asthenosphere, mesosphere, outer core, and inner core.
- Tectonic plates are large pieces of the lithosphere that move around on the Earth's surface.
- Knowledge about the structure of the Earth comes from the study of seismic waves caused by earthquakes.

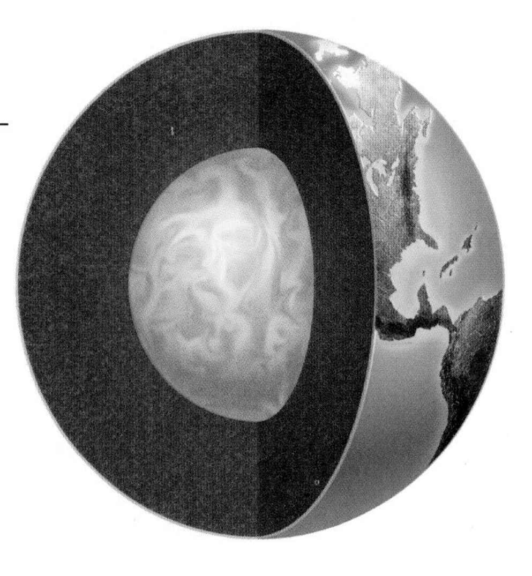

SECTION 2

Vocabulary

continental drift *(p. 95)*
sea-floor spreading *(p. 97)*

Section Notes

- Wegener's theory of continental drift explained many puzzling facts, including the fit of the Atlantic coastlines of South America and Africa.
- Today's continents were originally joined together in the ancient continent Pangaea.
- Some of the most important evidence for sea-floor spreading comes from magnetic reversals recorded in the ocean floor.

Skills Check

Math Concepts

MAKING MODELS Suppose you built a model of the Earth that had a radius of 100 cm (diameter of 200 cm). The radius of the real Earth is 6,378 km, and the thickness of its outer core is 2,200 km. What percentage of the Earth's radius is the outer core? How thick would the outer core be in your model?

$$\frac{2{,}200 \text{ km}}{6{,}378 \text{ km}} = 0.34 = 34\%$$

$$34\% \text{ of } 100 \text{ cm} = 0.34 \times 100 \text{ cm} = 34 \text{ cm}$$

Visual Understanding

SEA-FLOOR SPREADING This close-up view of a mid-ocean ridge shows how new oceanic lithosphere forms. As the two tectonic plates pull away from each other, magma fills in the cracks that open between them. When this magma solidifies, it becomes the newest part of the oceanic plate.

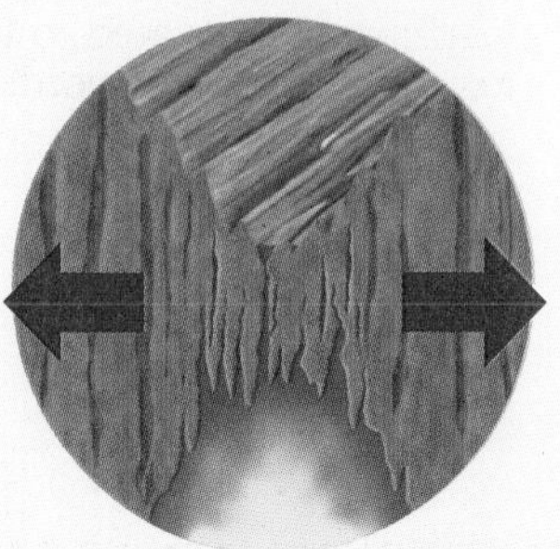

SECTION 3

Vocabulary

plate tectonics *(p. 99)*
convergent boundary *(p. 100)*
subduction zone *(p. 100)*
divergent boundary *(p. 101)*
transform boundary *(p. 101)*

Section Notes

- The processes of ridge push, convection, and slab pull provide some possible driving forces for plate tectonics.
- Tectonic plate boundaries are classified as convergent, divergent, or transform.
- Data from satellite tracking indicate that some tectonic plates move an average of 3 cm a year.

SECTION 4

Vocabulary

stress *(p. 103)*
compression *(p. 103)*
tension *(p. 103)*
folding *(p. 104)*
fault *(p. 105)*
normal fault *(p. 105)*
reverse fault *(p. 105)*
strike-slip fault *(p. 106)*

Section Notes

- As tectonic plates move next to and into each other, a great amount of stress is placed on the rocks at the boundary.
- Folding occurs when rock layers bend due to stress.
- Faulting occurs when rock layers break due to stress and then move on either side of the break.
- Mountains are classified as either folded, fault-block, or volcanic, depending on how they form.
- Mountain building is caused by the movement of tectonic plates. Different types of movement cause different types of mountains.

internetconnect

GO TO: go.hrw.com

Visit the **HRW** Web site for a variety of learning tools related to this chapter. Just type in the keyword:

KEYWORD: HSTTEC

GO TO: www.scilinks.org

Visit the **National Science Teachers Association** on-line Web site for Internet resources related to this chapter. Just type in the *sci*LINKS number for more information about the topic:

TOPIC: Composition of the Earth **_sci_LINKS NUMBER:** HSTE155
TOPIC: Structure of the Earth **_sci_LINKS NUMBER:** HSTE160
TOPIC: Tectonic Plates **_sci_LINKS NUMBER:** HSTE165
TOPIC: Faults **_sci_LINKS NUMBER:** HSTE170
TOPIC: Mountain Building **_sci_LINKS NUMBER:** HSTE175

Chapter Review

USING VOCABULARY

For each pair of terms, explain the difference in their meanings.

1. oceanic crust/continental crust
2. lithosphere/asthenosphere
3. convergent boundary/divergent boundary
4. folding/faulting
5. oceanic crust/oceanic lithosphere
6. normal fault/reverse fault

UNDERSTANDING CONCEPTS

Multiple Choice

7. The part of the Earth that is a liquid is the
   a. crust.
   b. mantle.
   c. outer core.
   d. inner core.

8. The part of the Earth on which the tectonic plates are able to move is the
   a. lithosphere.
   b. asthenosphere.
   c. mesosphere.
   d. subduction zone.

9. The ancient continent that contained all the landmasses is called
   a. Pangaea.
   b. Gondwana.
   c. Laurasia.
   d. Panthalassa.

10. The type of tectonic plate boundary involving a collision between two tectonic plates is
    a. divergent.
    b. transform.
    c. convergent.
    d. normal.

11. The type of tectonic plate boundary that sometimes has a subduction zone is
    a. divergent.
    b. transform.
    c. convergent.
    d. normal.

12. The San Andreas fault is an example of a
    a. divergent boundary.
    b. transform boundary.
    c. convergent boundary.
    d. normal boundary.

13. When a fold is shaped like an arch, with the fold in an upward direction, it is called a(n)
    a. monocline.
    b. anticline.
    c. syncline.
    d. decline.

14. The type of fault in which the hanging wall moves down relative to the footwall is called
    a. strike-slip.
    b. reverse.
    c. normal.
    d. fault-block.

15. The type of mountain involving huge sections of the Earth's crust being pushed up into anticlines and synclines is the
    a. folded mountain.
    b. fault-block mountain.
    c. volcanic mountain.
    d. strike-slip mountain.

16. Continental mountain ranges are usually associated with
    a. divergent boundaries.
    b. transform boundaries.
    c. convergent boundaries.
    d. normal boundaries.

17. Mid-ocean ridges are associated with
    a. divergent boundaries.
    b. transform boundaries.
    c. convergent boundaries.
    d. normal boundaries.

Short Answer

18. What is a tectonic plate?

19. What was the major problem with Wegener's theory of continental drift?

20. Why is there stress on the Earth's crust?

Concept Mapping

21. Use the following terms to create a concept map: sea-floor spreading, convergent boundary, divergent boundary, subduction zone, transform boundary, tectonic plates.

CRITICAL THINKING AND PROBLEM SOLVING

Write one or two sentences to answer each of the following questions:

22. Why is it necessary to think about the different layers of the Earth in terms of both their composition and their physical properties?

23. Folded mountains usually form at the edge of a tectonic plate. How can you explain old folded mountain ranges located in the middle of a tectonic plate?

24. New tectonic plate material continually forms at divergent boundaries. Tectonic plate material is also continually destroyed in subduction zones at convergent boundaries. Do you think the total amount of lithosphere formed on Earth is about equal to the amount destroyed? Why?

MATH IN SCIENCE

25. Assume that a very small oceanic plate is between a mid-ocean ridge to the west and a subduction zone to the east. At the ridge, the oceanic plate is growing at a rate of 5 km every million years. At the subduction zone, the oceanic plate is being destroyed at a rate of 10 km every million years. If the oceanic plate is 100 km across, in how many million years will the oceanic plate disappear?

INTERPRETING GRAPHICS

Imagine that you could travel to the center of the Earth. Use the diagram below to answer the questions that follow.

Composition	Structure
Crust (50 km)	Lithosphere (150 km)
Mantle (2,900 km)	Asthenosphere (250 km)
	Mesosphere (2,550 km)
Core (3,428 km)	Outer core (2,200 km)
	Inner core (1,228 km)

26. How far beneath Earth's surface would you have to go to find the liquid material in the Earth's core?

27. At what range of depth would you find mantle material but still be within the lithosphere?

Take a minute to review your answers to the Pre-Reading Questions found at the bottom of page 86. Have your answers changed? If necessary, revise your answers based on what you have learned since you began this chapter.

Science, Technology, and Society

Living on the Mid-Atlantic Ridge

Imagine living hundreds of kilometers from other people on an icy outcrop of volcanic rock surrounded by the cold North Atlantic Ocean. How would you stay warm? For the people of Iceland, this is an important question that affects their daily lives. Iceland is a volcanic island situated on the Mid-Atlantic Ridge, just south of the Arctic Circle. Sea-floor spreading produces active volcanoes, earthquakes, hot springs, and geysers that make life on this island seem a little unstable. However, the same volcanic force that threatens civilization provides the heat necessary for daily life. Icelanders use the geothermal energy supplied by their surroundings in ways that might surprise you.

▲ *The Blue Lagoon in Iceland is the result of producing energy from water power.*

Let's Go Geothermal!

Geothermal literally means "earth heat," *geo-* meaning "earth" and *therme* meaning "heat." Around the ninth century A.D., Iceland's earliest settlers took advantage of the Earth's heat by planting crops in naturally heated ground. This encouraged rapid plant growth and an early harvest of food. In 1928, Iceland built its first public geothermal utility project—a hole drilled into the Earth in order to pump water from a hot spring. After the oil crisis of the 1970s, geothermal-energy projects were built on a grand scale in Iceland. Today 85 percent of all houses in Iceland are heated by geothermal energy. Hot water from underground pools is pumped directly to houses, where it is routed through radiators to provide heating.

Geothermal water is also pumped to homes to provide hot tap water. This natural source meets all the hot-water needs for the city of Reykjavik, with a population of about 150,000 people!

There are still other uses for this hot water. For example, it is used to heat 120 public swimming pools. Picture yourself swimming outside in naturally hot water during the dead of winter! Greenhouses, where fruits and vegetables are grown, are also warmed by this water. Even fish farming on Iceland's exposed coastline wouldn't be possible without geothermal energy to adjust the water temperature. In other industries, geothermal energy is used to dry timber, wool, and seaweed.

Power Production

Although hydropower (producing energy from water power) is the principal source of electricity in Iceland, geothermal energy is also used. Water ranging in temperature from 300–700°C is pumped into a reservoir, where the water turns into steam that forces turbines to turn. The spinning motion of these turbines generates electricity. Power generation from geothermal sources is only about 5–15 percent efficient and results in a very large amount of water runoff. At the Svartsengi power plant, this water runoff has created a beautiful pool that swimmers call the Blue Lagoon.

Going Further

▶ Can you think of other abundant clean-energy resources? How could we harness such sources?

SCIENTIFICDEBATE

Continental Drift

When Alfred Wegener proposed his theory of continental drift in the early 1900s, many scientists laughed at the idea of continents plowing across the ocean. In fact, many people found his theory so ridiculous that Wegener, a university professor, had difficulty getting a job! Wegener's theory jolted the very foundation of geology.

Wegener's Theory

Wegener used geologic, fossil, and glacial evidence gathered on opposite sides of the Atlantic Ocean to support his theory of continental drift. For example, Wegener recognized geologic similarities between the Appalachian Mountains, in eastern North America, and the Scottish Highlands, as well as similarities between rock strata in South Africa and Brazil. He believed that these striking similarities could be explained only if these geologic features were once part of the same continent.

Wegener proposed that because they are less dense, continents float on top of the denser rock of the ocean floor. Although continental drift explained many of Wegener's observations, he could not find scientific evidence to develop a complete explanation of how continents move.

Alfred Wegener (1880–1930)

The Critics

Most scientists were skeptical of Wegener's theory and dismissed it as foolishness. Some critics held fast to old theories that giant land bridges could explain similarities among fossils in South America and Africa. Others argued that Wegener's theory could not account for the tremendous forces that would have been required to move continents such great distances. Wegener, however, believed that these forces could be the same forces responsible for earthquakes and volcanic eruptions.

The Evidence

During the 1950s and 1960s, discoveries of sea-floor spreading and magnetic reversal provided the evidence that Wegener's theory needed and led to the theory of plate tectonics. The theory of plate tectonics describes how the continents move. Today geologists recognize that continents are actually parts of moving tectonic plates that float on the asthenosphere, a layer of partially molten rock.

Like the accomplishments of so many scientists, Wegener's accomplishments went unrecognized until years after his death. The next time you hear a scientific theory that sounds far out, don't underestimate it. It may be proven true!

Also an Astronomer and Meteorologist

Wegener had a very diverse background in the sciences. He earned a Ph.D. in astronomy from the University of Berlin. But he was always very interested in geophysics and meteorology. His interest in geophysics led to his theory on continental drift. His interest in meteorology eventually led to his death. He froze to death in Greenland while returning from a rescue mission to bring food to meteorologists camped on a glacier.

On Your Own

▶ Photocopy a world map. Carefully cut out the continents from the map. Be sure to cut along the line where the land meets the water. Slide the continents together like a jigsaw puzzle. How does this relate to the tectonic plates and continental drift?

CHAPTER 5

Earthquakes

Sections

1. What causes earthquakes?
2. Why are some earthquakes stronger than others?
3. Why do some buildings remain standing during earthquakes while others fall down?

If You Build It, Will It Stand?

On September 21, 1999, the island of Taiwan was forever changed. At 1:47 A.M., an earthquake struck, toppling buildings and burying thousands of people in rubble. Why did this building collapse while those that surrounded it did not? The collapsed building was not built to be as strong as the other buildings. In this chapter, you will learn about what causes earthquakes and what you can do to prepare for one. You will also learn how buildings can be constructed to withstand the force of an earthquake.

Search and rescue dogs help save lives after an earthquake.

START-UP Activity

BEND, BREAK, OR SHAKE

If you were in a building during an earthquake, what would you want the building to be made of? To answer this question, you need to know how building materials react to stress.

Procedure

1. Gather a **small wooden stick,** a **wire clothes hanger,** and a **plastic clothes hanger.**
2. Draw a straight line on a **sheet of paper.** Use a **protractor** to measure and draw the following angles from the line: 20°, 45°, and 90°.
3. Put on your safety goggles. Using the angles that you drew as a guide, try bending each item 20° and then releasing it. What happens? Does it break? If it bends, does it return to its original shape? Write your observations in your ScienceLog.
4. Repeat step 3, but bend each item 45°. Repeat the test again, but bend each item 90°.

Analysis

5. How do the materials' responses to bending compare?
6. Where earthquakes happen, engineers use building materials that are flexible but do not break or stay bent. Which materials from this experiment would you want building materials to behave like? Explain your answer.

SECTION 1

READING WARM-UP

Terms to Learn

seismology
fault
deformation
elastic rebound
seismic waves
P waves
S waves

What You'll Do

- Determine where earthquakes come from and what causes them.
- Identify different types of earthquakes.
- Describe how earthquakes travel through the Earth.

What Are Earthquakes?

The word *earthquake* defines itself fairly well. But there is more to an earthquake than just ground shaking. In fact, there is a branch of Earth science devoted to earthquakes called seismology (siez MAHL uh jee). **Seismology** is the study of earthquakes. Earthquakes are complex, and they present many questions for *seismologists,* the scientists who study earthquakes.

Where Do Earthquakes Occur?

Most earthquakes take place near the edges of tectonic plates. *Tectonic plates* are giant masses of solid rock that make up the outermost part of the Earth. **Figure 1** shows the Earth's tectonic plates and the locations of recent major earthquakes recorded by scientists.

Tectonic plates move in different directions and at different speeds. Two plates can push toward each other or pull away from each other. They can also slip past each other like slow-moving trains traveling in opposite directions.

As a result of these movements, numerous features called faults exist in the Earth's crust. A **fault** is a break in the Earth's crust along which blocks of the crust slide relative to one another. Earthquakes occur along faults due to this sliding.

Faults occur in many places, but they are especially common near the edges of tectonic plates where they form the boundaries along which the plates move. This is why earthquakes are so common near tectonic plate boundaries.

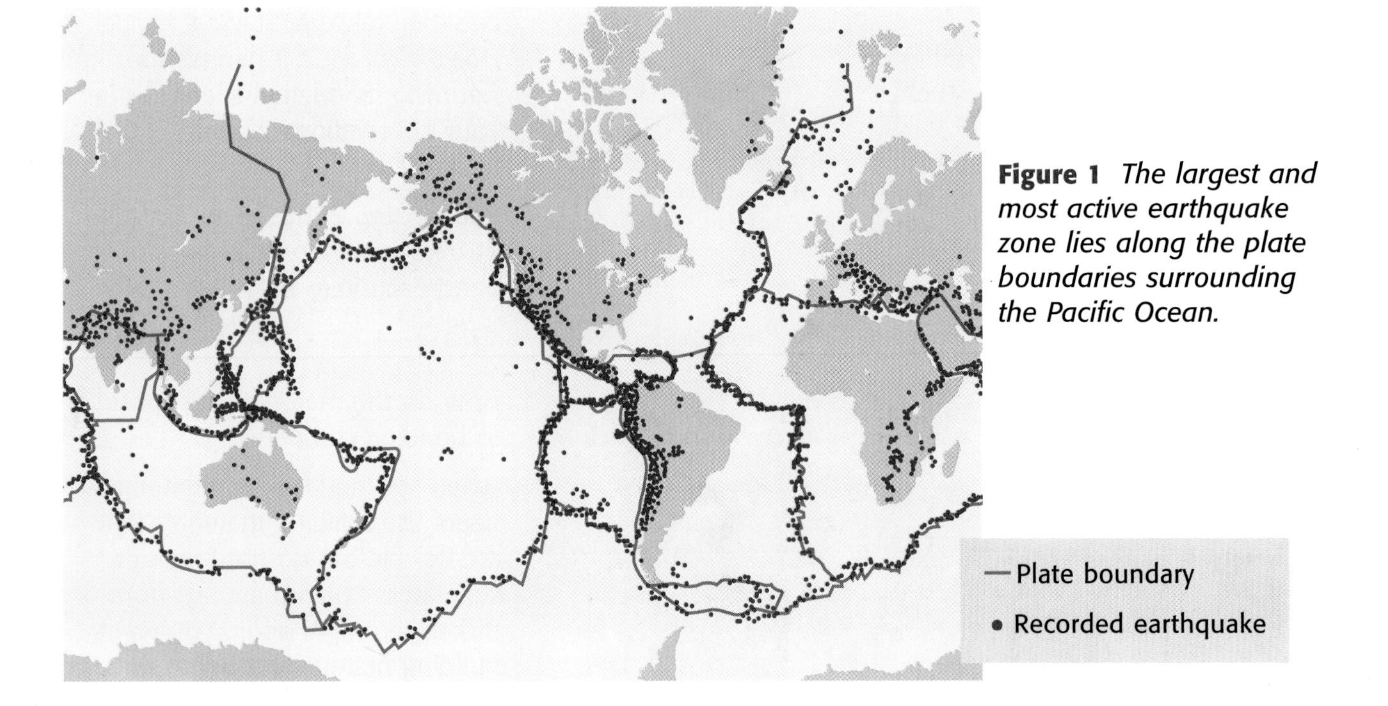

Figure 1 *The largest and most active earthquake zone lies along the plate boundaries surrounding the Pacific Ocean.*

What Causes Earthquakes?

As tectonic plates push, pull, or scrape against each other, stress builds up along faults near the plates' edges. In response to this stress, rock in the plates deforms. **Deformation** is the change in the shape of rock in response to stress. Rock along a fault deforms in mainly two ways—in a plastic manner, like a piece of molded clay, or in an elastic manner, like a rubber band. *Plastic deformation,* which is shown in **Figure 2,** does not lead to earthquakes.

Elastic deformation, however, does lead to earthquakes. While rock can stretch farther than steel without breaking, it will break at some point. Think of elastically deformed rock as a stretched rubber band. You can stretch a rubber band only so far before it breaks. When the rubber band breaks, it releases energy, and the broken pieces return to their unstretched shape.

Figure 2 *This photograph, taken in Hollister, California, shows how plastic deformation along the Calaveras Fault permanently bent a wall.*

Like the return of the broken rubber-band pieces to their unstretched shape, **elastic rebound** is the sudden return of elastically deformed rock to its original shape. Elastic rebound occurs when more stress is applied to rock than the rock can withstand. During elastic rebound, rock releases energy that causes an earthquake, as shown in **Figure 3.**

Figure 3 Elastic Rebound and Earthquakes

1 The rock along the fault has no stress acting on it.

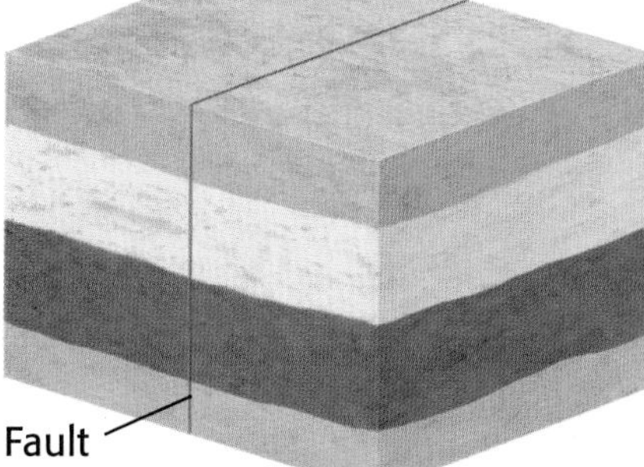

2 Tectonic forces push rock on either side of the fault in opposite directions, but the rock is locked together and does not move. The rock deforms in an elastic manner.

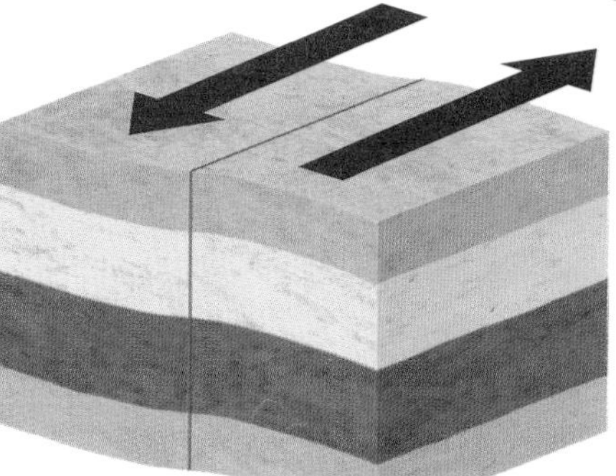

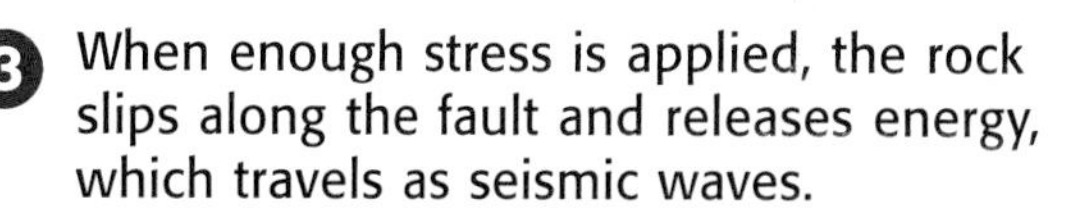

3 When enough stress is applied, the rock slips along the fault and releases energy, which travels as seismic waves.

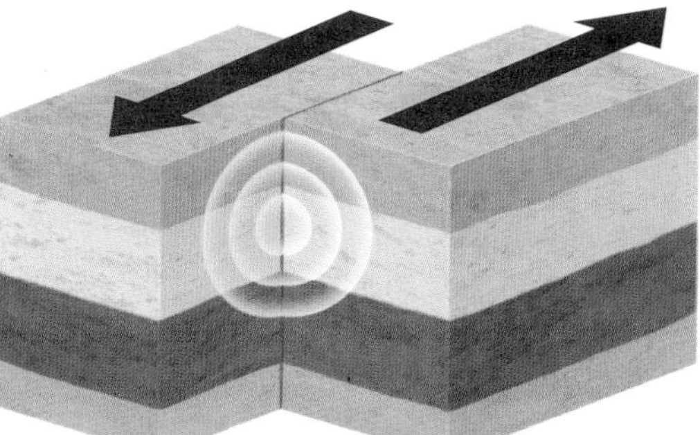

Are All Earthquakes the Same?

Earthquakes differ in strength and in the depth at which they begin. These differences depend on the type of tectonic plate motion that produces the earthquake. Examine the chart and the diagram below to learn how earthquakes differ.

Plate motion	Prominent fault type	Earthquake characteristics
Transform	strike-slip fault	moderate, shallow
Convergent	reverse fault	strong, deep
Divergent	normal fault	weak, shallow

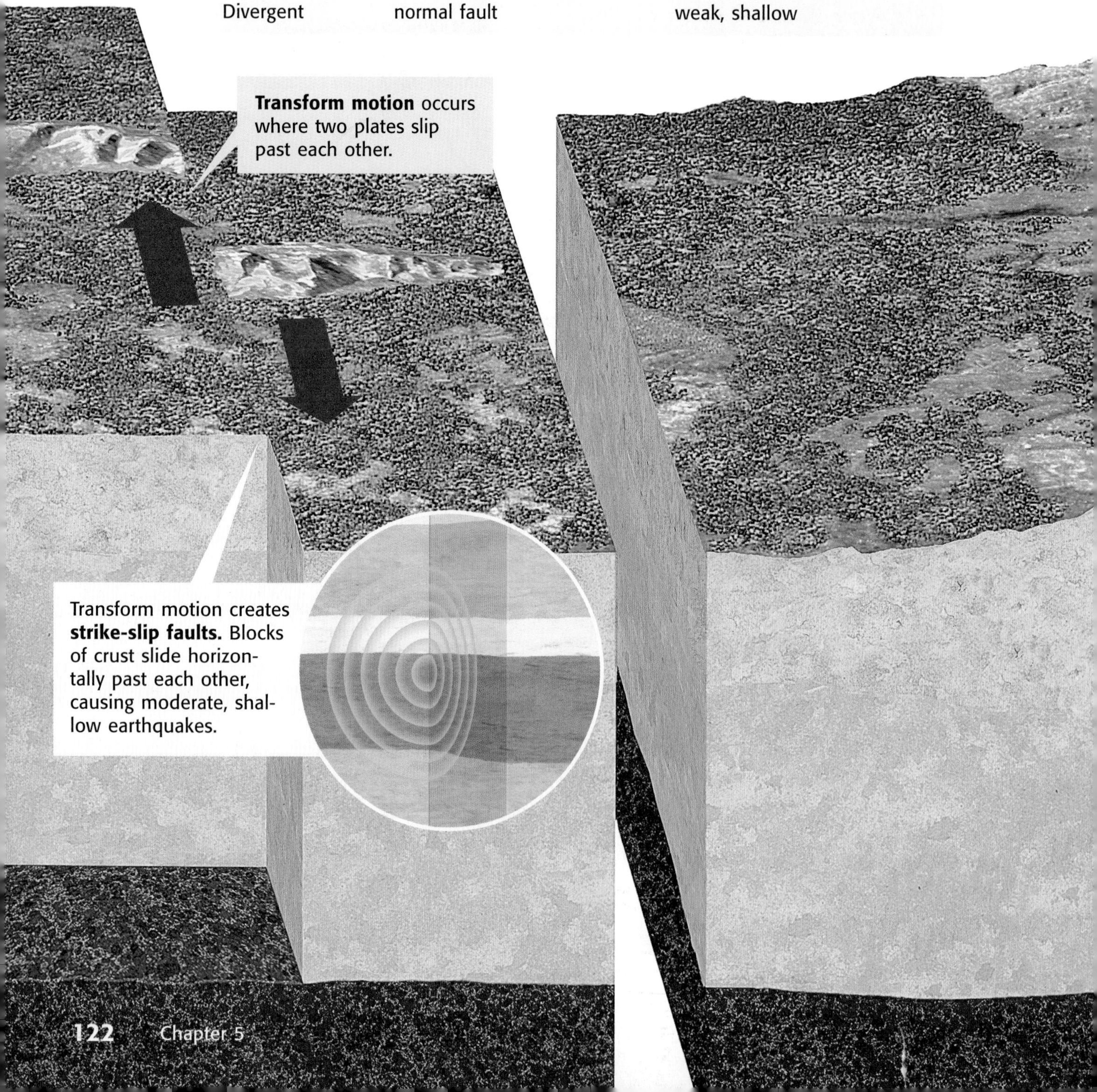

Transform motion occurs where two plates slip past each other.

Transform motion creates **strike-slip faults.** Blocks of crust slide horizontally past each other, causing moderate, shallow earthquakes.

Self-Check

Name two differences between the results of convergent motion and the results of divergent motion. *(See page 216 to check your answer.)*

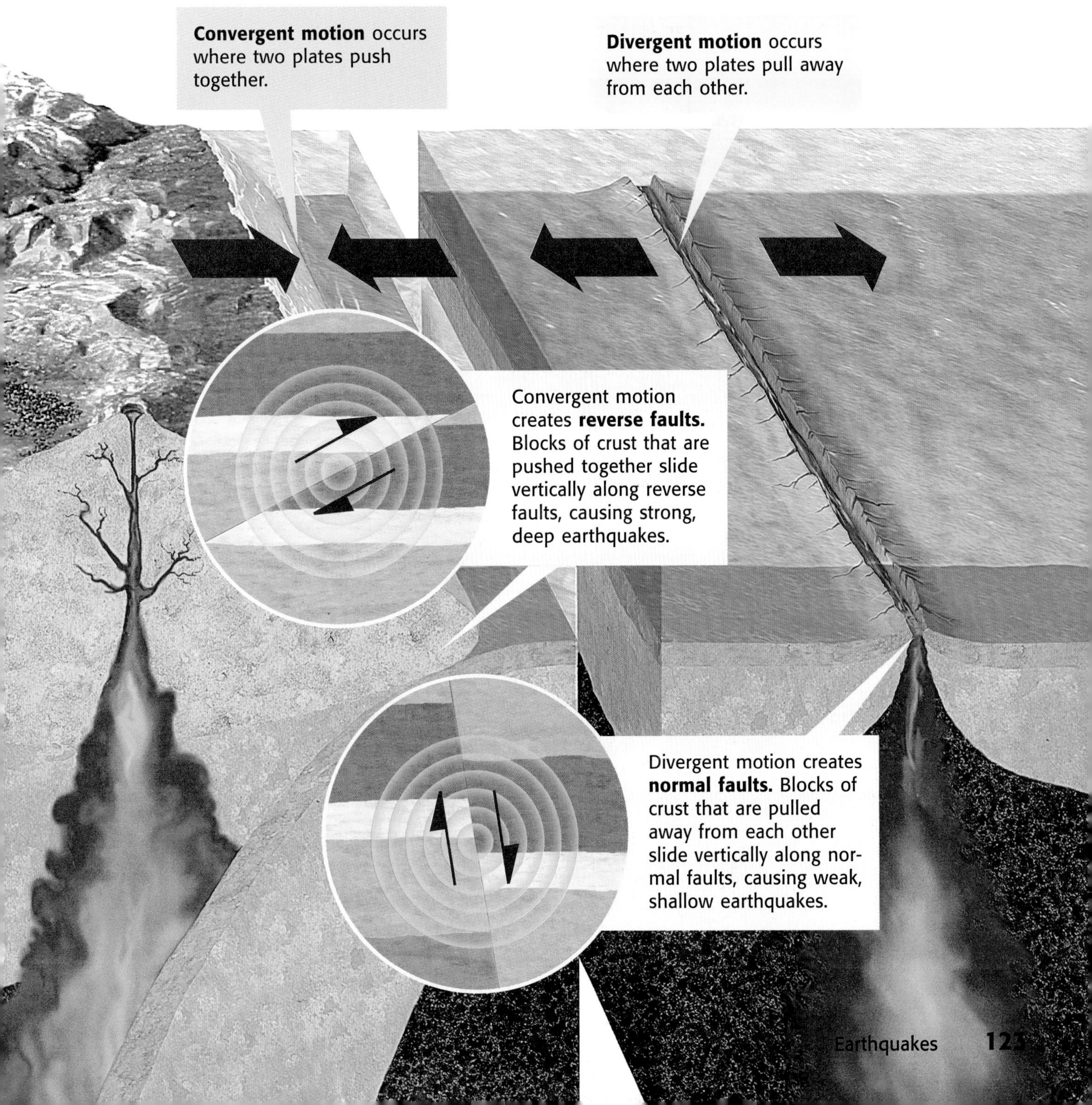

Convergent motion occurs where two plates push together.

Divergent motion occurs where two plates pull away from each other.

Convergent motion creates **reverse faults.** Blocks of crust that are pushed together slide vertically along reverse faults, causing strong, deep earthquakes.

Divergent motion creates **normal faults.** Blocks of crust that are pulled away from each other slide vertically along normal faults, causing weak, shallow earthquakes.

All types of waves share basic features. Understanding one type, such as seismic waves, can help you understand many other types. Other types of waves include light waves, sound waves, and water waves.

How Do Earthquakes Travel?

Remember that rock releases energy when it springs back after being deformed. This energy travels in the form of seismic waves. **Seismic waves** are waves of energy that travel through the Earth. Seismic waves that travel through the Earth's interior are called *body waves*. There are two types of body waves: P waves and S waves. Seismic waves that travel along the Earth's surface are called *surface waves*. Different types of seismic waves travel at different speeds and move the materials that they travel through differently.

Direction of wave travel

Figure 4 *P waves move rock back and forth between a squeezed position and a stretched position as they travel through it.*

P Is for Primary If you squeeze an elastic material into a smaller volume or stretch it into a larger volume, the pressure inside the material changes. When you suddenly stop squeezing or stretching the material, it springs briefly back and forth before returning to its original shape. This is how P waves (pressure waves) affect rock, as shown in **Figure 4. P waves,** which travel through solids, liquids, and gases, are the fastest seismic waves. Because they are the fastest seismic waves and because they can move through all parts of the Earth, P waves always travel ahead of other seismic waves. Because P waves are always the first seismic waves to be detected, they are also called *primary* waves.

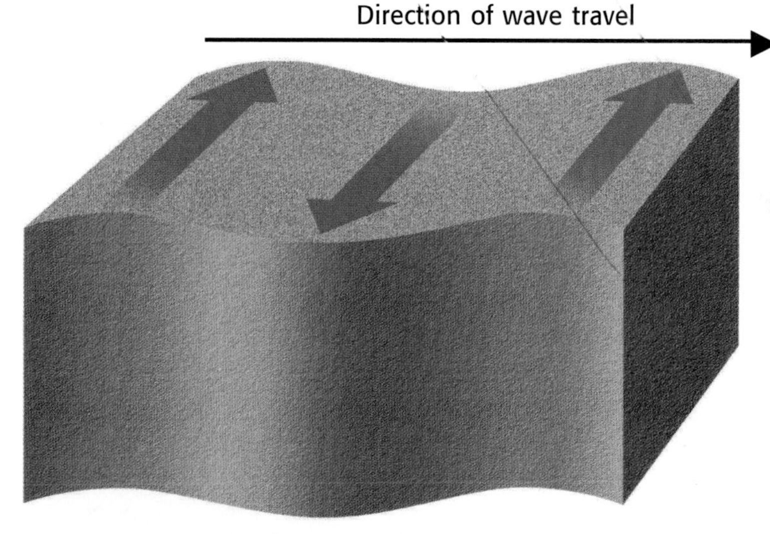

Figure 5 *S waves shear rock back and forth as they travel through it.*

S Is for Secondary Rock can also be deformed from side to side. When the rock springs back to its original position after being deformed, S waves are created. **S waves,** or shear waves, are the second-fastest seismic wave. S waves shear rock back and forth, as shown in **Figure 5.** *Shearing* stretches parts of rock sideways from other parts.

Unlike P waves, S waves cannot travel through parts of the Earth that are completely liquid. Also, S waves are slower than P waves and always arrive second; thus, they are also called *secondary* waves.

Surface Waves Surface waves move the ground up and down in circles as the waves travel along the surface. This is shown in **Figure 6.** Many people have reported feeling like they were on a roller coaster during an earthquake. This feeling comes from surface waves passing along the Earth's surface. Surface waves travel more slowly than body waves but are more destructive. Most damage during an earthquake comes from surface waves, which can literally shake the ground out from under a building.

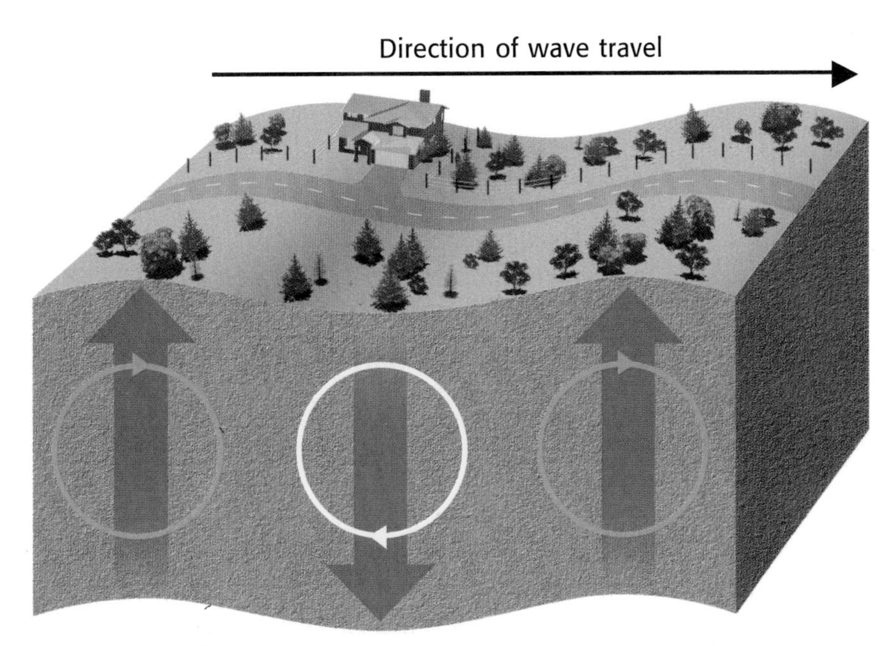

Figure 6 *Surface waves move the ground much like ocean waves move water particles.*

Modeling Seismic Waves

1. Stretch a **spring toy** lengthwise on a **table.**
2. Hold one end of the spring while a partner holds the other end. Push your end toward your partner's end, and observe what happens.
3. Repeat step 2, but this time shake the spring from side to side.
4. Which type of seismic wave is represented in step 2? in step 3?

SECTION REVIEW

1. Where do earthquakes occur?
2. What directly causes earthquakes?
3. Arrange the types of earthquakes caused by the three plate-motion types from weakest to strongest.
4. **Analyzing Relationships** Why are surface waves more destructive to buildings than P waves or S waves?

TOPIC: What Is an Earthquake?
GO TO: www.scilinks.org
***sci*LINKS NUMBER:** HSTE180

SECTION 2

READING WARM-UP

Earthquake Measurement

Terms to Learn

seismograph
seismogram
epicenter
focus

What You'll Do

- Explain how earthquakes are detected.
- Demonstrate how to locate earthquakes.
- Describe how the strength of an earthquake is measured.

After an earthquake occurs, seismologists try to find out when and where it started. Earthquake-sensing devices enable seismologists to record and measure seismic waves. These measurements show how far the seismic waves traveled. The measurements also show how much the ground moved. Seismologists use this information to pinpoint where the earthquake started and to find out how strong the earthquake was.

Locating Earthquakes

How do seismologists know when and where earthquakes begin? They depend on earthquake-sensing instruments called seismographs. **Seismographs** are instruments located at or near the surface of the Earth that record seismic waves. When the waves reach a seismograph, the seismograph creates a seismogram, such as the one in **Figure 7.** A **seismogram** is a tracing of earthquake motion created by a seismograph.

Figure 7 *The line in a seismogram traces the movement of the ground as it shakes. The more the ground moves, the farther back and forth the line traces.*

When Did It Happen? Seismologists use seismograms to calculate when an earthquake started. An earthquake starts when rock slips suddenly enough along a fault to create seismic waves. Seismologists find an earthquake's start time by comparing seismograms and noting the difference in arrival times of P waves and S waves.

Where Did It Happen? Seismologists also use seismograms to find an earthquake's epicenter. An **epicenter** is the point on the Earth's surface directly above an earthquake's starting point. A **focus** is the point inside the Earth where an earthquake begins. **Figure 8** shows the relationship between an earthquake's epicenter and its focus.

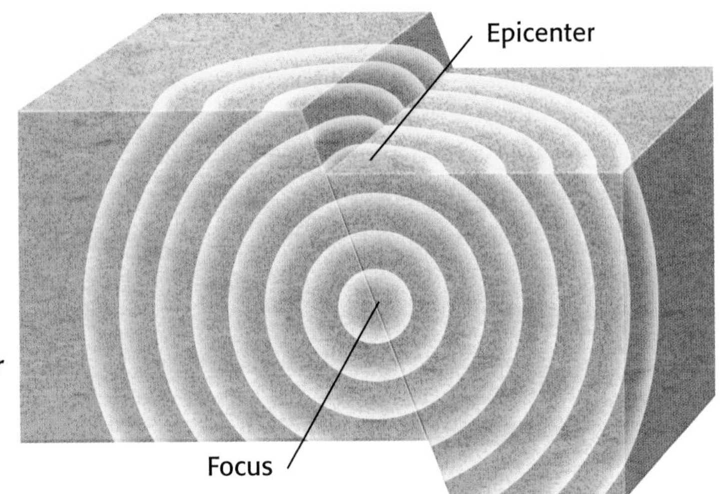

Figure 8 *An earthquake's epicenter is on the Earth's surface directly above the earthquake's focus.*

Putting It All Together Perhaps the most common method by which seismologists find an earthquake's epicenter is the *S-P-time method.* When using the S-P-time method, seismologists begin by collecting several seismograms of the same earthquake from different locations. Seismologists then place the seismograms on a time-distance graph so the first P waves line up with the P-wave curve and the first S waves line up with the S-wave curve. This is shown in **Figure 9.**

After the seismograms are placed on the graph, seismologists can see how far away from each station the earthquake was by reading the distance axis. After seismologists find out the distances, they can find the earthquake's epicenter as shown below.

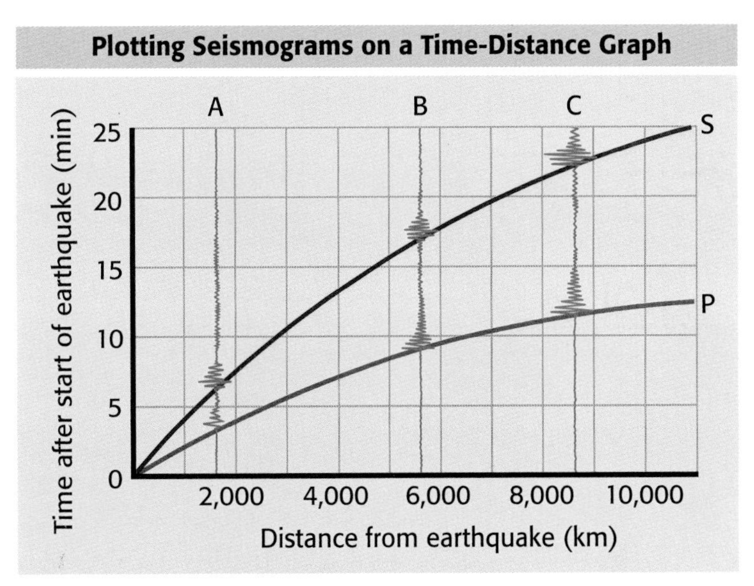

Figure 9 *Seismologists subtract a wave's travel time (read from the vertical axis) from the time that the wave was recorded. This indicates when the earthquake started. The distance of the stations from the epicenter is read from the horizontal axis.*

Finding an Earthquake's Epicenter

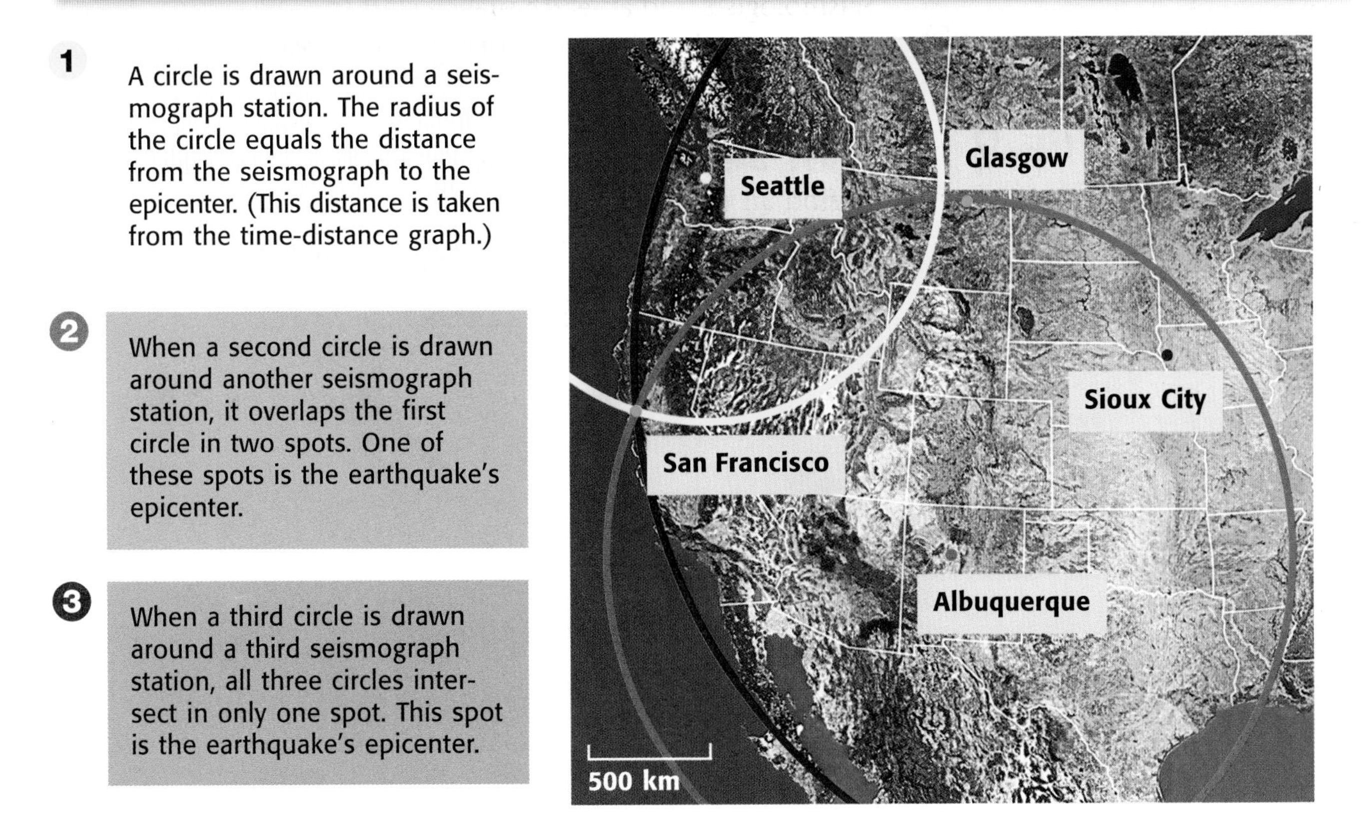

MATH BREAK

Moving Up the Scale

If the amount of energy released by an earthquake with a magnitude of 2.0 on the Richter scale is *n*, what are the amounts of energy released by earthquakes with the following magnitudes in terms of *n*: 3.0, 4.0, 5.0, and 6.0? (Hint: The energy released by an earthquake with a magnitude of 3.0 is 31.7*n*.)

Measuring Earthquake Strength

"How strong was the earthquake?" is a common question asked of seismologists. This is not an easy question to answer. But it is an important question for public officials, safety organizations, and businesses as well as seismologists. Fortunately, seismograms can be used not only to determine an earthquake's epicenter and its start time but also to find out an earthquake's strength.

The Richter Scale The *Richter scale* is commonly used to measure earthquake strength. It is named after Charles Richter, an American seismologist who developed the scale in the 1930s. A modified version of the Richter scale is shown below.

Modified Richter Scale

Magnitude	Estimated effects
2.0	can be detected only by seismograph
3.0	can be felt at epicenter
4.0	felt by most in area
5.0	causes damage at epicenter
6.0	causes widespread damage
7.0	causes great, widespread damage

Earthquake Energy There is a pattern in the Richter scale relating an earthquake's magnitude and the amount of energy released by the earthquake. Each time the magnitude increases by 1 unit, the amount of energy released becomes 31.7 times larger. For example, an earthquake with a magnitude of 5.0 on the Richter scale will release 31.7 times as much energy as an earthquake with a magnitude of 4.0 on the Richter scale.

internet**connect**

*sci*LINKS NSTA

TOPIC: Earthquake Measurement
GO TO: www.scilinks.org
***sci*LINKS NUMBER:** HSTE185

SECTION REVIEW

1. What is the difference between a seismogram and a seismograph?
2. How many seismograph stations are needed to use the S-P-time method? Why?
3. **Doing Calculations** If the amount of energy released by an earthquake with a magnitude of 7.0 on the Richter scale is *x*, what is the amount of energy released by an earthquake with a magnitude of 6.0 in terms of *x*?

SECTION 3

READING WARM-UP

Terms to Learn

gap hypothesis
seismic gap

What You'll Do

- Explain earthquake hazard.
- Compare methods of earthquake forecasting.
- List ways to safeguard buildings against earthquakes.
- Outline earthquake safety procedures.

Earthquakes and Society

Earthquakes are a fascinating part of Earth science, but they are very dangerous. Seismologists have had some success in predicting earthquakes, but simply being aware of earthquakes is not enough. It is important for people in earthquake-prone areas to be prepared.

Earthquake Hazard

Earthquake hazard measures how prone an area is to experiencing earthquakes in the future. An area's earthquake-hazard level is determined by past and present seismic activity. Look carefully at the map in **Figure 10.** As you can see, some areas of the United States have a higher earthquake-hazard level than others. This is because some areas have more seismic activity than others. The West Coast, for example, has a very high earthquake-hazard level because it has a lot of seismic activity. Areas such as the Gulf Coast or the Midwest have much lower earthquake-hazard levels because they do not have as much seismic activity.

Can you find the area where you live on the map? What level or levels of earthquake hazard are shown for your area? Look at the hazard levels in nearby areas. How do their hazard levels compare with your area's hazard level? What could explain the earthquake-hazard levels in your area and nearby areas?

Figure 10 *This is an earthquake-hazard map of the continental United States. It shows various levels of earthquake hazard for different areas of the country.*

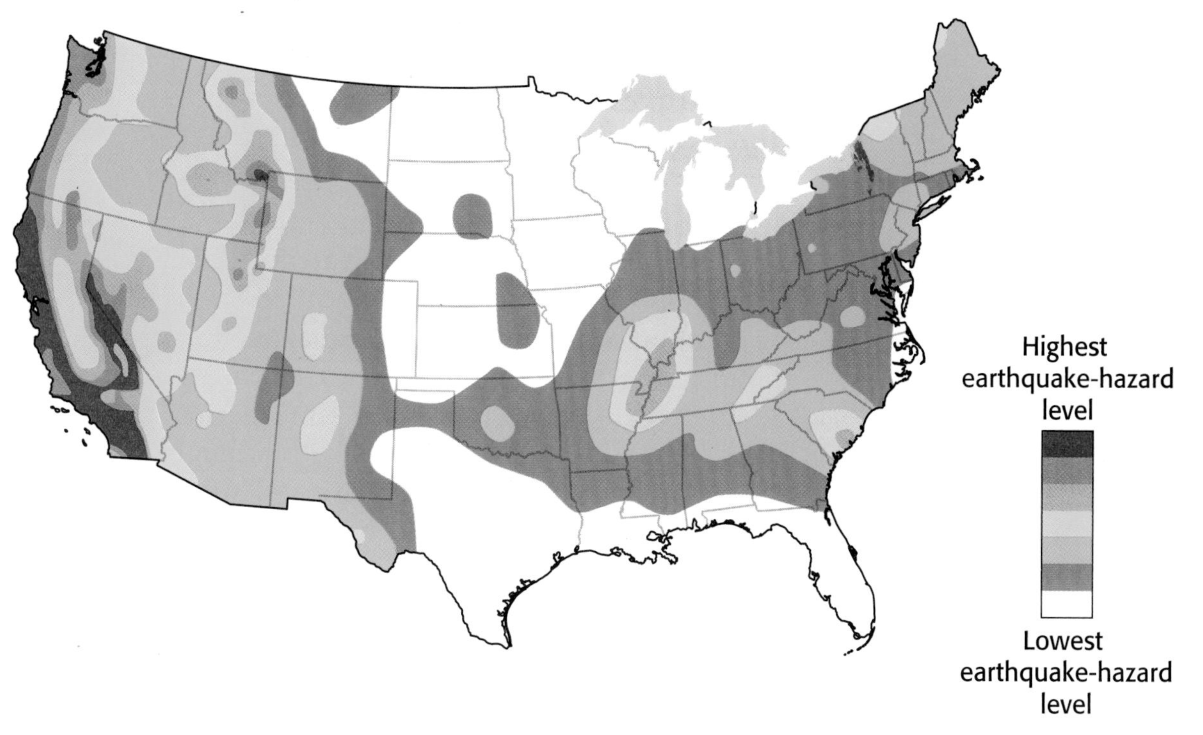

Earthquake Forecasting

Predicting when and where earthquakes will occur and how strong they will be is a difficult task. However, by closely monitoring active faults and other areas of seismic activity, seismologists have discovered some patterns in earthquakes that allow them to make some broad predictions.

Strength and Frequency As you learned earlier, earthquakes vary in strength. And you can probably guess that earthquakes don't occur on a set schedule. But what you may not know is that the strength of earthquakes is related to how often they occur. The chart in **Figure 11** provides more detail on this relationship.

Self-Check

According to the chart below, about how many earthquakes with a magnitude between 6.0 and 6.9 occur annually?
(See page 216 to check your answer.)

Figure 11 *Generally, with each step down in earthquake magnitude, the number of earthquakes per year is about 10 times greater.*

Worldwide Earthquake Frequency (Based on Observations Since 1900)		
Descriptor	**Magnitude**	**Average occurring annually**
Great	8.0 and higher	1
Major	7.0–7.9	18
Strong	6.0–6.9	120
Moderate	5.0–5.9	800
Light	4.0–4.9	about 6,200
Minor	3.0–3.9	about 49,000
Very minor	2.0–2.9	about 365,000

This relationship between earthquake strength and frequency is also observed on a local scale. For example, each year approximately 10 earthquakes occur in the Puget Sound area of Washington with a magnitude of 4 on the Richter scale. Over this same time period, approximately 10 times as many earthquakes with a magnitude of 3 occur in this area. Scientists use these statistics to make predictions about the strength, location, and frequency of future earthquakes.

Can animals predict earthquakes? To decide for yourself, turn to page 144 to read about links between animal behavior and earthquakes.

The Gap Hypothesis Another method of predicting an earthquake's strength, location, and frequency is based on the gap hypothesis. The **gap hypothesis** states that sections of active faults that have had relatively few earthquakes are likely to be the sites of strong earthquakes in the future. The areas along a fault where relatively few earthquakes have occurred are called **seismic gaps. Figure 12** below shows an example of a seismic gap.

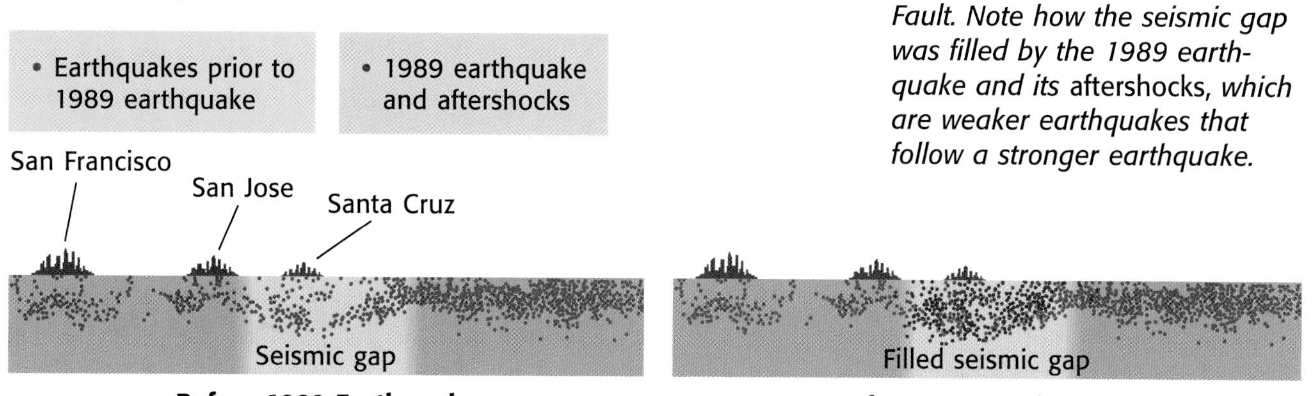

Figure 12 *This diagram shows a cross section of the San Andreas Fault. Note how the seismic gap was filled by the 1989 earthquake and its* aftershocks, *which are weaker earthquakes that follow a stronger earthquake.*

The gap hypothesis helped seismologists forecast the approximate time, strength, and location of the 1989 Loma Prieta earthquake in the San Francisco Bay area. The seismic gap that they identified is illustrated in Figure 12. In 1988, seismologists predicted that over the next 30 years there was a 30 percent chance that an earthquake with a magnitude of at least 6.5 would fill this seismic gap. Were they correct? The Loma Prieta earthquake, which filled in the seismic gap in 1989, measured 7.1 on the Richter scale. That's very close, considering how complicated the forecasting of earthquakes is.

Earthquakes and Buildings

Much like a judo master knocks the feet out from under his or her opponent, earthquakes shake the ground out from under buildings and bridges. Once the center of gravity of a structure has been displaced far enough off the structure's supporting base, most structures simply collapse.

Figure 13 shows what can happen to buildings during an earthquake. These buildings were not designed or constructed to withstand the forces of an earthquake.

Figure 13 *An earthquake shook the ground floor out from under the second story of this apartment building, which then collapsed.*

Activity

Research a tall building to find out how its structure is reinforced. Would any of the building's reinforcements safeguard it against earthquakes? Has an earthquake occurred in the building's area since the building was constructed? If so, how well did the building withstand the shaking?

TRY at HOME

Earthquake Resistant Buildings People have learned a lot from building failure during earthquakes. Architects and engineers use the newest technology to design and construct buildings and bridges to better withstand earthquakes. Study this diagram carefully to learn about some of this modern technology.

The **mass damper** is a weight placed in the roof of a building. Motion sensors detect building movement during an earthquake and send messages to a computer. The computer then signals controls in the roof to shift the mass damper to counteract the building's movement.

Steel **cross-braces** are placed between floors. These braces counteract pressure that pushes and pulls at the side of a building during an earthquake.

The **active tendon system** works much like the mass damper system in the roof. Sensors notify a computer that the building is moving. Then the computer activates devices to shift a large weight to counteract the movement.

Flexible pipes help prevent water and gas lines from breaking. Engineers design the pipes with flexible joints so the pipes are better able to twist and bend without breaking during an earthquake.

Base isolators act as shock absorbers during an earthquake. They are made of layers of rubber and steel wrapped around a lead core. Base isolators absorb seismic waves, preventing them from traveling through the building.

Are You Prepared for an Earthquake?

If you live in an earthquake-prone area or ever plan to visit one, there are many things you can do to protect yourself and your property from earthquakes. Plan ahead so you will know what to do before, during, and after an earthquake. Stick to your plan as closely as possible.

Before the Shaking Starts The first thing you should do is safeguard your house against earthquakes. For example, put heavier objects on lower shelves so they do not fall on anyone during the earthquake. You can also talk to adults about having your home reinforced. Make a plan with others (your family, neighbors, or friends) to meet somewhere after the earthquake is over. This way someone will know you are safe. During the earthquake, waterlines, power lines, and roadways may be damaged. Therefore, you should store nonperishable food, water, a fire extinguisher, a flashlight with batteries, and a first-aid kit in a place you can access after the earthquake.

Figure 14 *Simple precautions can greatly reduce the chance of injury during an earthquake.*

When the Shaking Starts The best thing to do if you are indoors is to crouch or lie face down under a table or desk in the center of a room, as shown in **Figure 15.** If you are outside, lie face down away from buildings, power lines, and trees, and cover your head with your hands. If you are in a car on an open road, you should stop the car and remain inside.

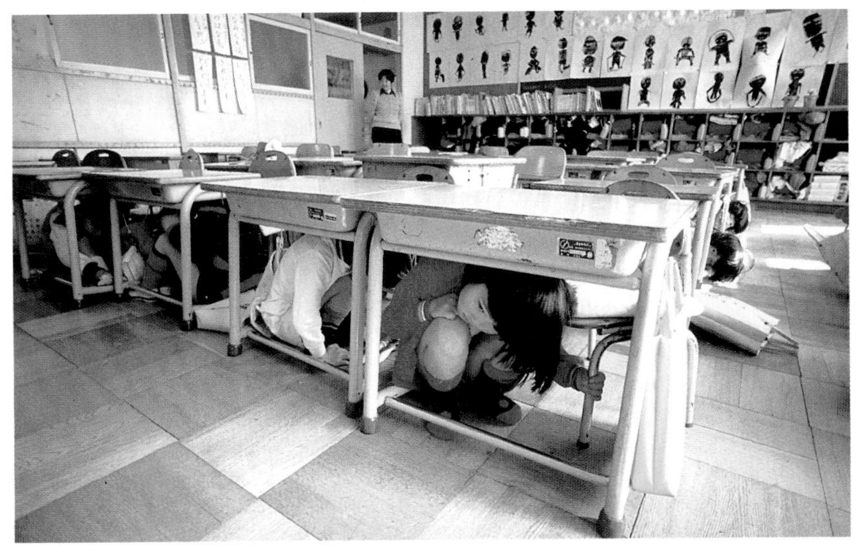

Figure 15 *These students are participating in an earthquake drill.*

internet**connect**

SC*i*LINKS. NSTA

TOPIC: Earthquakes and Society
GO TO: www.scilinks.org
***sci*LINKS NUMBER:** HSTE190

After the Shaking Stops Being in an earthquake is a startling experience. Afterward, you should not be surprised to find yourself and others puzzled about what happened. You should try to calm down, get your bearings, and remove yourself from immediate danger, such as downed power lines, broken glass, and fire hazards. Be aware that there may be aftershocks. Recall your earthquake plan, and follow it through.

SECTION REVIEW

1. How is an area's earthquake hazard determined?
2. Which earthquake forecast predicts a more precise location—a forecast based on the relationship between strength and frequency or a forecast based on the gap hypothesis?
3. Describe two ways that buildings are reinforced against earthquakes.
4. Name four items that you should store in case of an earthquake.
5. **Using Graphics** Would the street shown in the photo at left be a safe place during an earthquake? Why or why not?

Earthquake Safety Plan

You are at home reading the evening news. On the front page you read a report from the local seismology station. Scientists predict an earthquake in your area sometime in the near future. You realize that you are not prepared.

Make a detailed outline of how you would prepare yourself and your home for an earthquake. Then write a list of safety procedures to follow during an earthquake. When you are done, exchange your work with a classmate. How do your plans differ from your classmate's? How might you work together to improve your earthquake safety plans?

SECTION 4

READING WARM-UP

Terms to Learn

Moho
shadow zone

What You'll Do

- Describe how seismic studies reveal Earth's interior.
- Summarize seismic discoveries on other cosmic bodies.

Earthquake Discoveries Near and Far

The study of earthquakes has led to many important discoveries about the Earth's interior. Seismologists learn about the Earth's interior by observing how seismic waves travel through the Earth. Likewise, seismic waves on other cosmic bodies allow seismologists to study the interiors of those bodies.

Discoveries in Earth's Interior

Have you ever noticed how light bends in water? If you poke part of a pencil into water and look at it from a certain angle, the pencil looks bent. This is because the light waves that bounce off the pencil bend as they pass through the water's surface toward your eye. Seismic waves bend in much the same way as they travel through rock. Seismologists have learned a lot about the Earth's interior by studying how seismic waves bend.

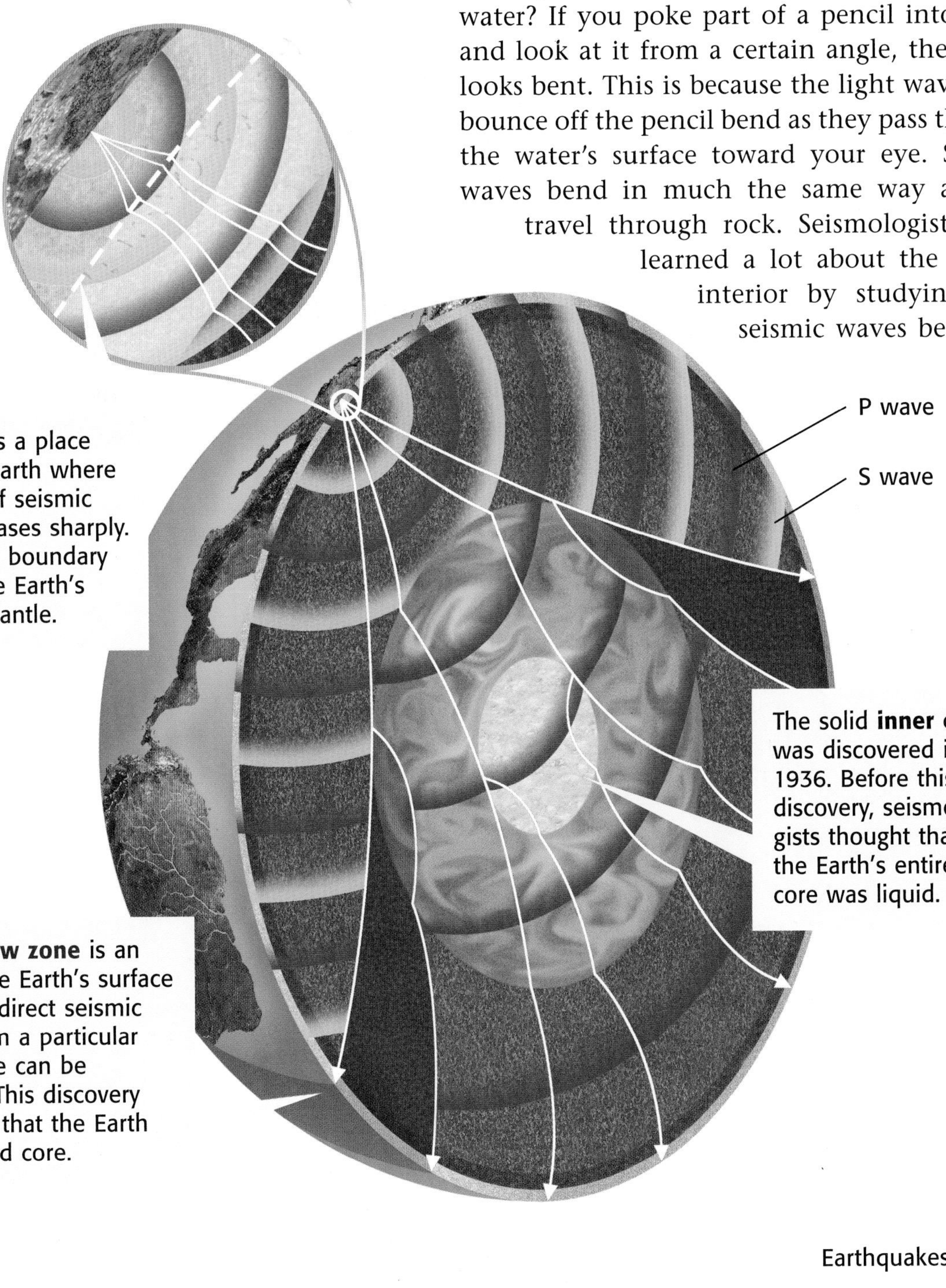

The **Moho** is a place within the Earth where the speed of seismic waves increases sharply. It marks the boundary between the Earth's crust and mantle.

The solid **inner core** was discovered in 1936. Before this discovery, seismologists thought that the Earth's entire core was liquid.

The **shadow zone** is an area on the Earth's surface where no direct seismic waves from a particular earthquake can be detected. This discovery suggested that the Earth has a liquid core.

Quakes and Shakes on Other Cosmic Bodies

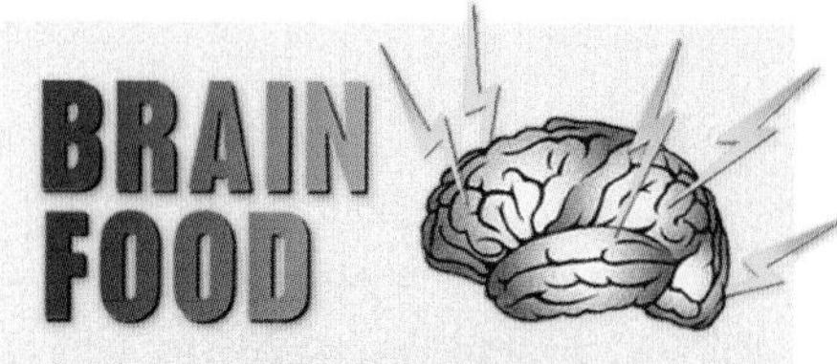

Many scientists think part of the moon was once part of the Earth. It is thought that when the Earth was almost entirely molten, a Mars-sized object collided with the Earth, knocking off part of Earth's mantle. The mantle material and material from the impacting body then began orbiting the Earth. Eventually, the orbiting material joined to form the moon.

Seismologists have taken what they have learned from earthquakes and applied it to studies of other cosmic bodies, such as planets, moons, and stars. They have been able to learn about the interiors of these cosmic bodies by studying how seismic waves behave within them. The first and perhaps most successful seismic test on another cosmic body was on Earth's moon.

The Moon In July 1969, humans set foot on the moon for the first time. They brought with them a seismograph. Not knowing if the moon was seismically active, they left nothing to chance—they purposely crashed their landing vehicle back into the moon's surface after they left to create artificial seismic waves. What happened after that left seismologists astonished.

If the lander had crashed into the Earth, the equivalent seismograms would have lasted 20–30 seconds at most. The surface of the moon, however, vibrated for more than an hour and a half! At first scientists thought the equipment was not working properly. But the seismograph recorded similar signals produced by meteoroid impacts and "moonquakes" long after the astronauts had left the moon. **Figure 16** shows the nature of these seismic events, which were observed remotely from Earth.

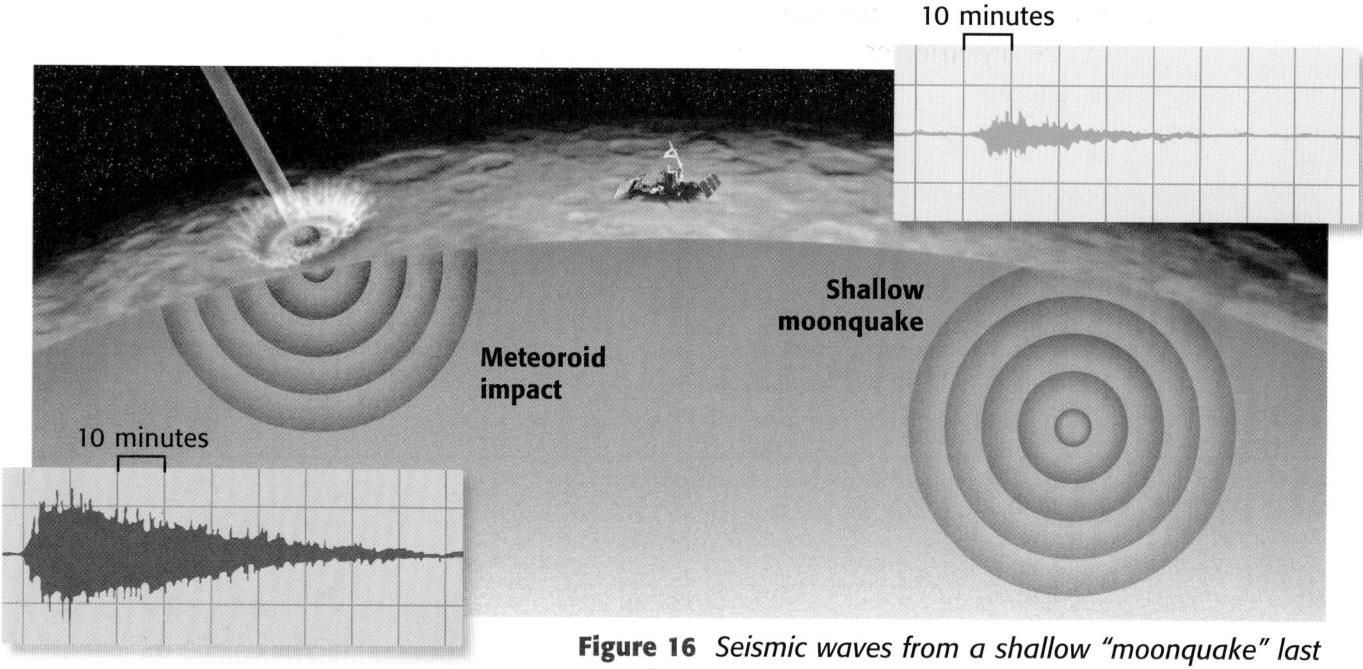

Figure 16 *Seismic waves from a shallow "moonquake" last 50 minutes. Seismic waves from a meteoroid impact last an hour and a half. Similar disturbances on Earth last less than a minute.*

Mars In 1976, a space probe called *Viking 1* allowed seismologists to learn about seismic activity on Mars. The probe, which was controlled remotely from Earth, landed on Mars and conducted several experiments. A seismograph was placed on top of the spacecraft to measure seismic waves on Mars. However, as soon as the craft landed, *Viking 1*'s seismograph began to shake. Scientists immediately discovered that Mars is a very windy planet and that the seismograph was working mainly as a wind gauge!

Although the wind on Mars interfered with the seismograph, the seismograph recorded seismograms for months. During that time, only one possible "marsquake" shook the seismograph harder than the wind did.

The Sun Seismologists have also studied seismic waves on the sun. Because humans cannot directly access the sun, scientists study it remotely by using a satellite called *SOHO*. Information gathered by *SOHO* has shown that solar flares produce seismic waves. *Solar flares* are powerful magnetic disturbances in the sun. The seismic waves that result cause "sunquakes," which are similar to earthquakes but are generally much stronger. For example, a moderate sunquake, shown in **Figure 17** beneath an image of *SOHO,* released more than 1 million times as much energy as the Great Hanshin earthquake mentioned at the beginning of this chapter!

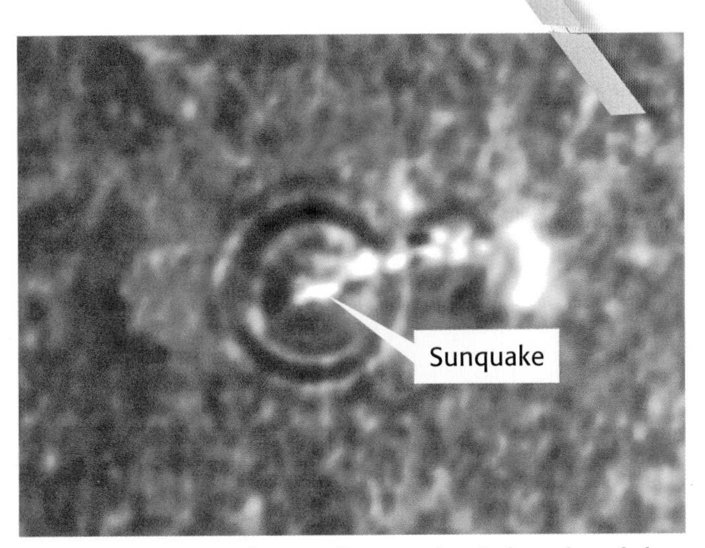

Figure 17 SOHO *detects "sunquakes" that dwarf the greatest earthquakes in history.*

SECTION REVIEW

1. What observation of seismic-wave travel led to the discovery of the Moho?
2. Briefly describe one discovery seismologists have made about each of the following cosmic bodies: the moon, Mars, and the sun.
3. **Interpreting Graphics** Take another look at the figure on the first page of Section 4. Why don't S waves enter the Earth's outer core?

Design Your Own Lab

USING SCIENTIFIC METHODS

Quake Challenge

In many parts of the world, it is important that buildings be built with earthquakes in mind. Each building must be planned so that the structure is protected during an earthquake. Architects have improved the design of buildings greatly since 1906, when an earthquake destroyed much of San Francisco. In this activity, you will use marshmallows and toothpicks to build a structure that can withstand a simulated earthquake. In the process, you will discover some of the ways that a building can be built to withstand an earthquake.

MATERIALS

- 10 marshmallows
- 10 toothpicks
- square of gelatin, approximately 8 cm × 8 cm
- paper plate

Ask a Question

1. What features help a building withstand an earthquake? How can I use this information to build my structure?

Form a Hypothesis

2. Brainstorm with a classmate to design a structure that will resist a simulated earthquake. Sketch your design in your ScienceLog. Write two or three sentences to describe your design. Explain why you think your design will be able to withstand a simulated earthquake.

Test the Hypothesis

3. Follow your design to build a structure using the toothpicks and marshmallows.
4. Place the gelatin square on the paper plate and set your structure on the gelatin.
5. Shake the square of gelatin to test whether your building will remain standing during an earthquake. Do not pick up the gelatin.
6. If your first design does not work well, change it until you find a design that does. Each time, try to determine why your building falls so that you can improve your design.

7. Sketch your final design in your ScienceLog.
8. After you have tested your final design, place your structure on the gelatin square on your teacher's desk.

9. When every group has added a structure to the teacher's gelatin, your teacher will simulate an earthquake by shaking the gelatin. Watch to see which buildings withstand the most severe earthquake.

Analyze the Results

10. Which buildings were still standing after the final earthquake? What features made them more stable?

11. How would you change your design to make your structure more stable?

Communicate Results

12. Based on this activity, what advice would you give to architects who design buildings in earthquake zones?

13. What are some limitations of your earthquake model?

14. How could your research have an impact on society?

Chapter Highlights

SECTION 1

Vocabulary

seismology *(p. 120)*
fault *(p. 120)*
deformation *(p. 121)*
elastic rebound *(p. 121)*
seismic waves *(p. 124)*
P waves *(p. 124)*
S waves *(p. 124)*

Section Notes

- Earthquakes mainly occur along faults near the edges of tectonic plates.
- Elastic rebound is the direct cause of earthquakes.
- Earthquakes differ depending on what type of plate motion causes them.
- Seismic waves are classified as body waves or surface waves.
- Body waves travel through the Earth's interior, while surface waves travel along the surface.
- There are two types of body waves: P waves and S waves.

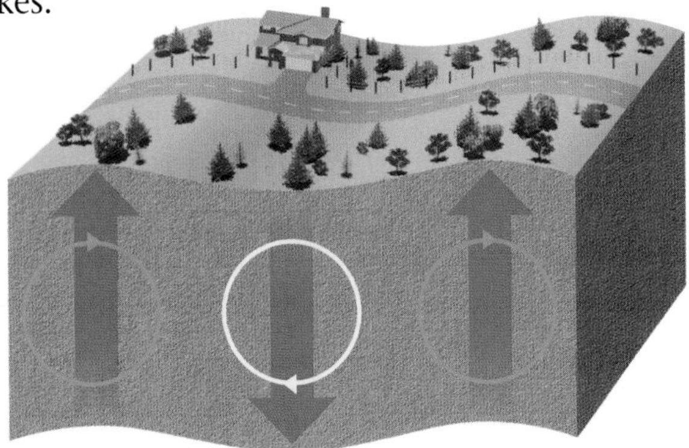

SECTION 2

Vocabulary

seismograph *(p. 126)*
seismogram *(p. 126)*
epicenter *(p. 126)*
focus *(p. 126)*

Section Notes

- Seismographs detect seismic waves and record them as seismograms.
- An earthquake's focus is the underground location where seismic waves begin. The earthquake's epicenter is on the surface directly above the focus.
- Seismologists use the S-P-time method to find an earthquake's epicenter.
- Seismologists use the Richter scale to measure an earthquake's strength.

Labs

Earthquake Waves *(p. 184)*

☑ Skills Check

Math Concepts

EARTHQUAKE STRENGTH The energy released by an earthquake increases by a factor of 31.7 with each increase in magnitude. The energy released decreases by a factor of 31.7 with each decrease in magnitude. All you have to do is multiply or divide.

If magnitude 4 releases energy y, then:

- magnitude 5 releases energy $31.7y$
- magnitude 3 releases energy $\frac{y}{31.7}$

Visual Understanding

TIME-DISTANCE GRAPH Note on the time-distance graph in Figure 9 that the difference in arrival times between P waves and S waves increases with distance from the epicenter.

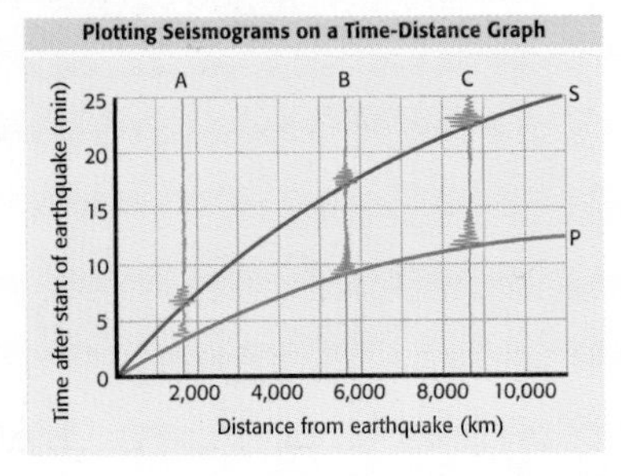

SECTION 3

Vocabulary

gap hypothesis *(p. 131)*
seismic gap *(p. 131)*

Section Notes

- Earthquake hazard measures how prone an area is to experiencing earthquakes in the future.
- Some earthquake predictions are based on the relationship between earthquake strength and earthquake frequency. As earthquake frequency decreases, earthquake strength increases.

- Predictions based on the gap hypothesis target seismically inactive areas along faults for strong earthquakes in the future.
- An earthquake usually collapses a structure by displacing the structure's center of gravity off the structure's supporting base.
- Buildings and bridges can be reinforced to minimize earthquake damage.
- People in earthquake-prone areas should plan ahead for earthquakes.

SECTION 4

Vocabulary

Moho *(p. 135)*
shadow zone *(p. 135)*

Section Notes

- The Moho, shadow zone, and inner core are features discovered on and inside Earth by observing seismic waves.
- Seismology has been used to study other cosmic bodies.
- Seismic waves last much longer on the moon than they do on Earth.
- Based on early seismic studies, Mars appears much less active seismically than the Earth.
- "Sunquakes" produce energy far greater than any earthquakes we know of.

internetconnect

GO TO: go.hrw.com

Visit the **HRW** Web site for a variety of learning tools related to this chapter. Just type in the keyword:

KEYWORD: HSTEQK

GO TO: www.scilinks.org

Visit the **National Science Teachers Association** on-line Web site for Internet resources related to this chapter. Just type in the ***sci*LINKS** number for more information about the topic:

TOPIC: What Is an Earthquake? — ***sci*LINKS NUMBER:** HSTE180
TOPIC: Earthquake Measurement — ***sci*LINKS NUMBER:** HSTE185
TOPIC: Earthquakes and Society — ***sci*LINKS NUMBER:** HSTE190
TOPIC: Earthquake Discoveries Near and Far — ***sci*LINKS NUMBER:** HSTE195

Chapter Review

USING VOCABULARY

To complete the following sentences, choose the correct term from each pair of terms listed below:

1. Energy is released as __?__ occurs. *(deformation* or *elastic rebound)*

2. __?__ cannot travel through parts of the Earth that are completely liquid. *(S waves* or *P waves)*

3. Seismic waves are recorded by a __?__. *(seismograph* or *seismogram)*

4. Seismologists use the S-P-time method to find an earthquake's __?__. *(shadow zone* or *epicenter)*

5. The __?__ is a place that marks a sharp increase in seismic wave speed. *(seismic gap* or *Moho)*

UNDERSTANDING CONCEPTS

Multiple Choice

6. When rock is __?__, energy builds up in it. Seismic waves occur as this energy is __?__.
   - **a.** elastically deformed; released
   - **b.** plastically deformed; released
   - **c.** elastically deformed; increased
   - **d.** plastically deformed; increased

7. The strongest earthquakes usually occur
   - **a.** near divergent boundaries.
   - **b.** near convergent boundaries.
   - **c.** near transform boundaries.
   - **d.** along normal faults.

8. The last seismic waves to arrive are
   - **a.** P waves.
   - **b.** S waves.
   - **c.** surface waves.
   - **d.** body waves.

9. If an earthquake begins while you are in a building, the safest thing to do first is
   - **a.** get under the strongest table, chair, or other piece of furniture.
   - **b.** run out into the street.
   - **c.** crouch near a wall.
   - **d.** call home.

10. Studying earthquake waves currently allows seismologists to do all of the following *except*
    - **a.** determine when an earthquake started.
    - **b.** learn about the Earth's interior.
    - **c.** decrease an earthquake's strength.
    - **d.** determine where an earthquake started.

11. If a planet has a liquid core, then S waves
    - **a.** speed up as they travel through the core.
    - **b.** maintain their speed as they travel through the core.
    - **c.** change direction as they travel through the core.
    - **d.** cannot pass through the core.

Short Answer

12. What is the relationship between the strength of earthquakes and earthquake frequency?

13. You learned earlier that if you are in a car during an earthquake and are out in the open, it is best to stay in the car. Briefly describe a situation in which you might want to leave a car during an earthquake.

14. How did seismologists determine that the outer core of the Earth is liquid?

Concept Mapping

15. Use the following terms to create a concept map: focus, epicenter, earthquake start time, seismic waves, P waves, S waves.

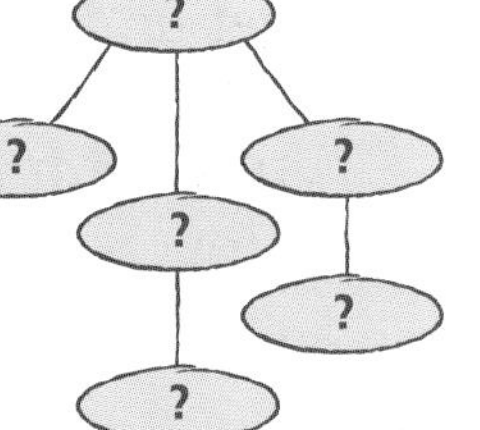

CRITICAL THINKING AND PROBLEM SOLVING

Write one or two sentences to answer the following questions:

16. How might the wall in Figure 2 appear if it had deformed elastically instead of plastically?

17. Why do strong earthquakes occur where there have not been many recent earthquakes? (Hint: Think about what gradually happens to rock before an earthquake occurs.)

18. What could be done to solve the wind problem with the seismograph on Mars? Explain how you would set up the seismograph.

MATH IN SCIENCE

19. Based on the relationship between earthquake magnitude and frequency, if 150 earthquakes with a magnitude of 2 occur in your area this year, about how many earthquakes with a magnitude of 4 should occur in your area this year?

INTERPRETING GRAPHICS

The graph below illustrates the relationship between earthquake magnitude and the height of the tracings on a seismogram. Charles Richter initially formed his magnitude scale by comparing the heights of seismogram readings for different earthquakes. Study the graph, and then answer the questions that follow.

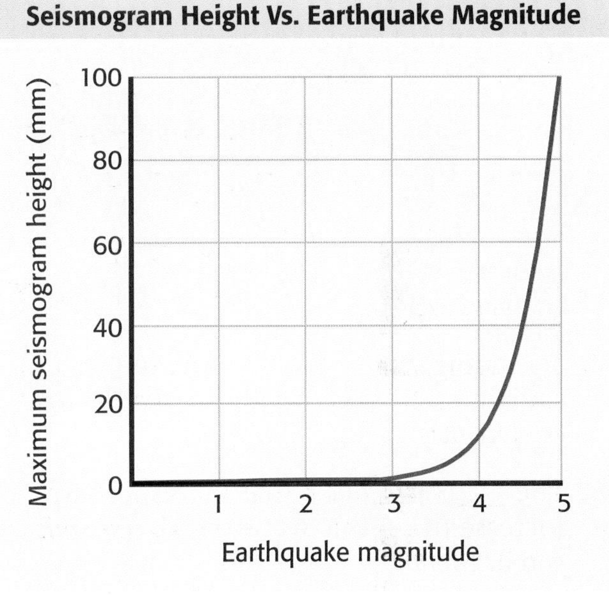

20. What would the magnitude of an earthquake be if the height of its seismogram readings were 10 mm?

21. Look at the shape of the curve on the graph. What does this tell you about the relationship between seismogram heights and earthquake magnitudes? Explain.

Take a minute to review your answers to the Pre-Reading Questions found at the bottom of page 118. Have your answers changed? If necessary, revise your answers based on what you have learned since you began this chapter.

WEIRD SCIENCE

CAN ANIMALS PREDICT EARTHQUAKES?

It Could Happen to You!

One day you come home from visiting a friend for the weekend and learn that your dog Pepper is hiding under your bed. Your father explains that he has been trying to get Pepper out from under the bed for the last 6 hours. Just then your mother enters the room and says that she has found two snakes in the backyard—and that makes a total of five in 2 days! This is very odd because you usually don't find more than one each year.

All the animals seem to be acting very strange. Your goldfish is even hiding behind a rock. You wonder if there is some explanation.

What's Going On?

So what's your guess? What do you think is happening? Did you guess that an earthquake is about to occur? Well, if you did, you are probably right!

Publications from as far back as 1784 record unusual animal behavior prior to earthquakes. Some examples included zoo animals refusing to go into their shelters at night and domestic cattle seeking high ground. Other animals, like lizards, snakes, and small mammals, evacuate their underground burrows, and wild birds leave their usual habitats. All of these events occurred a few days, several hours, or a few minutes before the earthquakes happened.

Animals on Call?

Today the majority of scientists look to physical instruments in order to help them predict earthquakes. Yet the fact remains that none of the geophysical instruments we have allow scientists to predict exactly when an earthquake will occur. Could animals know the answer?

▼ *Goldfish or earthquake sensor?*

There are changes in the Earth's crust that occur prior to an earthquake, such as magnetic field changes, subsidence (sinking), tilting, and bulging of the surface. These things can be monitored by modern instruments. Many studies have shown that electromagnetic fields affect the behavior of living organisms. Is it possible that animals close to the epicenter of an earthquake are able to sense changes in their environment? Should we pay attention?

You Decide

▶ Currently, the United States government does not fund research that investigates whether animals can predict earthquakes. Have a debate with your classmates about whether the government should fund such research.

EYE ON THE ENVIRONMENT

What Causes Such Destruction?

At 5:04 P.M. on October 14, 1989, life in California's San Francisco Bay Area seemed as normal as ever. The third game of the World Series was underway in Candlestick Park, now called 3Com Park. While 62,000 fans filled the park, other people were rushing home from a day's work. By 5:05 P.M., however, things had changed drastically. The fact sheet of destruction looks like this:

Injuries:	3,757
Deaths:	68
Damaged homes:	23,408
Destroyed homes:	1,018
Damaged businesses:	3,530
Destroyed businesses:	366
Financial loss:	over $6 billion

The Culprit

The cause of such destruction was a 7.1 magnitude earthquake that lasted for 20 seconds. Its epicenter was 97 km south of San Francisco in an area called Loma Prieta. The earthquake was so strong that people in San Diego and western Nevada (740 km away) felt it too. Considering the earthquake's high magnitude and the fact that it occurred during rush hour, it is amazing that more people did not die. However, the damage to buildings was widespread—it covered an area of 7,770 km^2. And by October 1, 1990, there had been more than 7,000 aftershocks of this quake.

Take Heed

Engineers and seismologists had expected a major earthquake, so the amount of damage they saw from this earthquake was no surprise. But experts agree that if the earthquake were of a higher magnitude or centered closer to Oakland, San Jose, or San Francisco, the damage would have been much worse. They are concerned that people who live in these areas aren't paying attention to the warning this earthquake represents.

Many people have a false sense of security because their buildings withstood the quake with little or no damage. But engineers and seismologists agree that the only reason the buildings survived was because the ground motion in those areas was fairly low.

Tomorrow May Be Too Late

Many buildings that withstood this earthquake were poorly constructed and would not withstand another earthquake. Experts say there is a 50 percent chance that one or more 7.0 magnitude earthquakes will occur in the San Francisco Bay Area in the next 30 years. And the results of the next quake could be much more devastating if people don't reinforce their buildings before it's too late.

▲ *Notice the different levels of destruction for various buildings on the same street.*

On Your Own

▶ Research the engineering innovations for constructing bridges and buildings in areas with seismic activity. Share your information with the class.

CHAPTER 6

Volcanoes

Sections

1. What causes a volcanic eruption?
2. What is lava, and how does it form?

Hot Lava, Quiet Eruption

Volcanic eruptions come in all sizes. In places like Hawaii, most eruptions are nonviolent. Lava flows in Hawaii are made of rock called basalt, which flows easily. Basaltic lava flows travel slowly but can reach a temperature of nearly 1,200°C! The lava flow shown here is slowly creeping across a road. As you can see, calm eruptions of lava can threaten property more than human life. In this chapter you will learn about nonexplosive eruptions, explosive eruptions, the formation of magma, and the ways that scientists are trying to predict volcanic eruptions.

This type of eruption is called a lava fountain.

ANTICIPATION

As you will see in this activity, volcanic eruptions are very difficult to predict.

Procedure

1. Place **10 mL of baking soda** in the center of a sheet of **bathroom tissue.** Fold the corners over the baking soda and crease the edges so that they stay in place. Place the tissue packet in the middle of a **large pan.**
2. Put **modeling clay** around the top edge of a **funnel.** Turn the funnel upside down over the tissue packet. Press down to make a tight seal.
3. Put your safety goggles on and add **50 mL of vinegar** and **several drops of liquid dish soap** to a 200 mL **beaker,** and stir.
4. Predict how much time will elapse before your volcano erupts.
5. Pour the liquid into the upturned funnel. Using a **stopwatch,** record the time you began to pour and the time your volcano erupts. How close was your prediction?

Analysis

6. How does your model represent the natural world?
7. What are some limitations of your model?
8. Based on the predictions of the entire class, what can you conclude about the accuracy of predicting volcanic eruptions?

SECTION 1

READING WARM-UP

Terms to Learn

volcano
lava
pyroclastic material

What You'll Do

- Distinguish between nonexplosive and explosive volcanic eruptions.
- Explain how the composition of magma determines the type of volcanic eruption that will occur.
- Classify the main types of lava and volcanic debris.

Volcanic Eruptions

Think about the force of the explosion produced by the first atomic bomb used in World War II. Now imagine an explosion 10,000 times stronger, and you get an idea of how powerful a volcanic eruption can be. As you may know, volcanic eruptions give rise to volcanoes. A **volcano** is a mountain that forms when molten rock, called *magma,* is forced to the Earth's surface.

Fortunately, few volcanoes give rise to explosive eruptions. Most eruptions are of a nonexplosive variety. You can compare these two types of eruptions by looking at the photographs on this and the next page.

Nonexplosive Eruptions

When people think of volcanic eruptions, they often imagine rivers of red-hot lava, called *lava flows.* Lava flows come from nonexplosive eruptions. **Lava** is magma that flows onto the Earth's surface. Relatively calm outpourings of lava, like the ones shown below, can release a huge amount of molten rock. Some of the largest mountains on Earth grew from repeated lava flows over hundreds of thousands of years.

▲ *Sometimes nonexplosive eruptions can spray lava into the air. Lava fountains, such as this one, rarely exceed a few hundred meters in height.*

In this nonexplosive eruption, a continuous stream of lava pours quietly from the crater of Kilauea, in Hawaii. ▶

◀ *Lava can flow many kilometers before it finally cools and hardens. As you can see in this photograph, lava flows often pose a greater threat to property than to human life.*

Explosive Eruptions

Take a look at **Figure 1.** In an explosive volcanic eruption, clouds of hot debris and gases shoot out from the volcano, often at supersonic speeds. Instead of producing lava flows, molten rock is blown into millions of pieces that harden in the air. The dust-sized particles can circle the globe for years in the upper atmosphere, while larger pieces of debris fall closer to the volcano.

In addition to shooting molten rock into the air, an explosive eruption can blast millions of tons of solid rock from a volcano. In a matter of minutes, an explosive eruption can demolish rock formations that took thousands of years to accumulate. Thus, as shown in **Figure 2,** a volcano may actually shrink in size rather than grow from repeated eruptions.

Figure 1 *In what resembles a nuclear explosion, volcanic debris rockets skyward during an eruption of Mount Redoubt, in Alaska.*

Figure 2 *Within minutes, the 1980 eruption of Mount St. Helens, in Washington, blasted away a whole side of the mountain, flattening and scorching 600 km^2 of forest.*

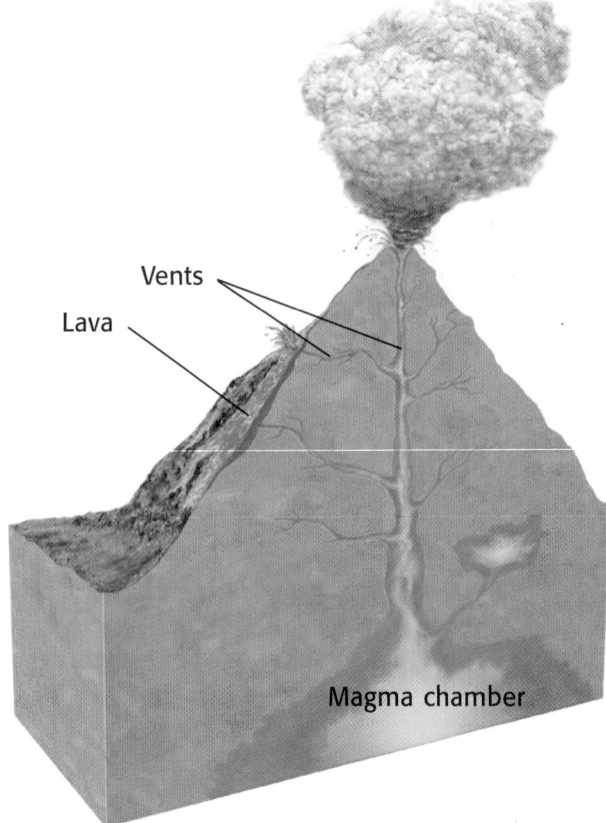

Cross Section of a Volcano

Whether they produce explosive or nonexplosive eruptions, all volcanoes share the same basic features. **Figure 3** shows some of the features that you might see if you could look inside an erupting volcano. Deep underground, the driving force that creates volcanoes is hot liquid material known as magma. Magma rises through holes in the Earth's crust called *vents*. Vents can channel magma all the way up to the Earth's surface during an eruption.

Figure 3 *Volcanoes form around vents that release magma onto the Earth's surface.*

Magma

By comparing the composition of magma from different types of eruptions, scientists have made an important discovery—the composition of the magma determines whether a volcanic eruption is nonexplosive, explosive, or somewhere in between.

Water A volcano is more likely to erupt explosively if its magma has a high water content. The effect water has on magma is similar to the effect carbon dioxide gas has in a can of soda. When you shake the can up, the carbon dioxide that was dissolved in the soda is released, and because gases need much more room than liquids, a great amount of pressure builds up. When you open the can, soda comes shooting out. The same phenomenon occurs with explosive volcanic eruptions.

Silica Explosive eruptions are also caused by magma that contains a large percentage of silica (a basic building block of most minerals). Silica-rich magma has a thick, stiff consistency. It flows slowly and tends to harden in the volcano's vent. This plugs the vent, resulting in a buildup of pressure as magma pushes up from below. If enough pressure builds up, an explosive eruption results. Thick magma also prevents water vapor and other gases from easily escaping. Magma that contains a smaller percentage of silica has a thinner, runnier consistency. Gases escape this type of magma more easily, making it less likely that explosive pressure will build up.

QuickLab

Bubble, Bubble, Toil and Trouble

With a few simple items, you can easily discover how the consistency of a liquid affects the flow of gases.

1. Fill a **drinking cup** halfway with **water** and another cup halfway with **honey.**
2. Using a **straw,** blow into the water and observe the bubbles.
3. Take another straw and blow into the honey. What happens?
4. How does the honey behave differently from the water?
5. How do you think this difference relates to volcanic eruptions?

TRY at HOME

What Erupts from a Volcano?

Depending on how explosive a volcanic eruption is, magma erupts as either *lava* or *pyroclastic material.* **Pyroclastic material** consists of the rock fragments created by explosive volcanic eruptions. Nonexplosive eruptions produce mostly lava. Explosive eruptions produce mostly pyroclastic material. Over many years, a volcano may alternate between eruptions of lava and eruptions of pyroclastic material. Eruptions of lava and pyroclastic material may also occur as separate stages of a single eruption event.

Fire and ice! A phrase to describe volcanoes? That depends on where they are. Turn to page 169 to find out more.

Lava Like magma, lava ranges in consistency from thick to thin. *Blocky lava* is so thick in consistency that it barely creeps along the ground. Other types of lava, such as *pahoehoe* (pah HOY HOY), *aa* (AH ah), and *pillow lava,* are thinner in consistency and produce faster lava flows. These types of lava are shown in the photographs below.

Blocky lava *is cool, stiff lava that cannot travel far from the erupting vent. Blocky lava usually oozes from a volcano, forming jumbled heaps of sharp-edged chunks.*

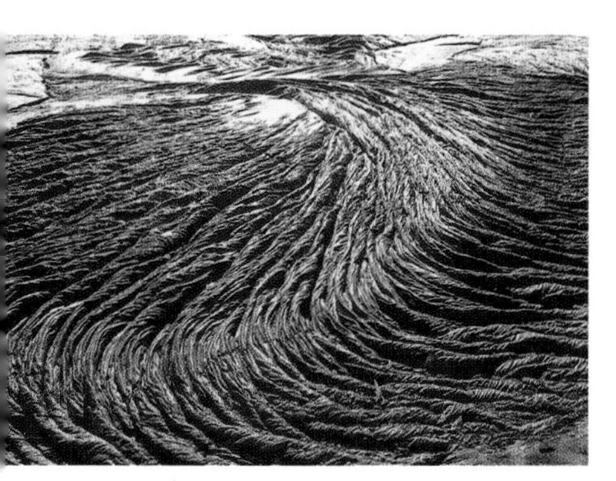

Pahoehoe *lava flows slowly, like wax dripping from a candle, forming a glassy surface with rounded wrinkles.*

Aa *is a Hawaiian word that refers to a type of lava that has a jagged surface. This slightly stiffer lava pours out quickly and forms a brittle crust. The crust is torn into jagged pieces as the molten lava underneath continues to move.*

Pillow lava *forms when lava erupts underwater. As you can see here, it forms rounded lumps that are the size and shape of pillows.*

Pyroclastic Material Pyroclastic material is produced when magma explodes from a volcano and solidifies in the air. It is also produced when existing rock is shattered by powerful eruptions. It comes in a variety of sizes, from boulders the size of houses to particles so small they can remain suspended in the atmosphere for years. The photographs on this page show four major kinds of pyroclastic material: volcanic bombs, volcanic blocks, lapilli (luh PILL ee), and volcanic ash.

Volcanic blocks *are the largest pieces of pyroclastic material. They consist of solid rock blasted out of the volcano.*

Volcanic bombs *are large blobs of magma that harden in the air. The shape of the bomb shown here resulted from the magma's spinning through the air as it cooled.*

Biology CONNECTION

Volcanoes provide some of the most productive farmland in the world. It can take thousands of years for volcanic rock to break down into usable soil nutrients. On the other hand, the ash from a single explosive eruption can greatly increase the fertility of soil in only a few years and can keep the soil fertile for centuries.

Lapilli, *which means "little stones" in Italian, are pebble-like bits of magma that became solid before they hit the ground.*

Volcanic ash *forms when the gases in stiff magma expand rapidly and the walls of the gas bubbles explode into tiny glasslike slivers.*

internet**connect**

sciLINKS NSTA

TOPIC: Volcanic Eruptions
GO TO: www.scilinks.org
***sci*LINKS NUMBER:** HSTE205

SECTION REVIEW

1. Is a nonexplosive volcanic eruption more likely to produce lava or pyroclastic material? Explain.
2. If a volcano contained magma with small proportions of water and silica, would you predict a nonexplosive eruption or an explosive one? Why?
3. **Making Inferences** Pyroclastic material is classified primarily by the size of the particles. What is the basis for classifying lava?

SECTION 2

READING WARM-UP

Volcanoes' Effects on Earth

Terms to Learn

shield volcano
cinder cone volcano
composite volcano
crater
caldera

What You'll Do

- Describe the effects that volcanoes have on Earth.
- Compare the different types of volcanoes.

The effects of volcanic eruptions can be seen both on land and in the air. Heavier pyroclastic materials fall to the ground, causing great destruction, while ash and escaping gases affect global climatic patterns. Volcanoes also build mountains and plateaus that become lasting additions to the landscape.

An Explosive Impact

Because it is thrown high into the air, ash ejected during explosive volcanic eruptions can have widespread effects. The ash can block out the sun for days over thousands of square kilometers. Volcanic ash can blow down trees and buildings and can blanket nearby towns with a fine powder.

Flows and Fallout As shown in **Figure 4,** clouds of hot ash can flow rapidly downhill like an avalanche, choking and searing every living thing in their path. Sometimes large deposits of ash mix with rainwater or the water from melted glaciers during an eruption. With the consistency of wet cement, the mixture flows downhill, picking up boulders, trees, and buildings along the way. As volcanic ash falls to the ground, the effects can be devastating. Buildings may collapse under the weight of so much ash. Ash can also dam up river valleys, resulting in massive floods. And although ash is an effective plant fertilizer, too much ash can smother crops, causing food shortages and loss of livestock.

Figure 4 *During the 1991 eruption of Mount Pinatubo, in the Philippines, clouds of volcanic gases and ash sped downhill at up to 250 km/h.*

Climatic Changes In large-scale eruptions, volcanic ash, along with sulfur-rich gases, can reach the upper atmosphere. As the ash and gases spread around the globe, they can block out enough sunlight to cause the average global surface temperature to drop noticeably. The eruption of Mount Pinatubo in 1991 caused average global temperatures to drop by as much as 0.5°C. Although this may not seem like a large change in temperature, such a shift can disrupt climates all over the world. The lower average temperatures may last for several years, bringing wetter, milder summers and longer, harsher winters.

Different Types of Volcanoes

The lava and pyroclastic material that erupt from volcanoes create a variety of landforms. Perhaps the best known of all volcanic landforms are the volcanoes themselves. Volcanoes result from the buildup of rock around a vent. Three basic types of volcanoes are illustrated in **Figure 5.**

Figure 5 Three Types of Volcanoes

Shield volcano

Shield volcanoes are built out of layers of lava from repeated nonexplosive eruptions. Because the lava is very runny, it spreads out over a wide area. Over time, the layers of lava create a volcano with gently sloping sides. Although their sides are not very steep, shield volcanoes can be enormous. Hawaii's Mauna Kea, the shield volcano shown here, is the largest mountain on Earth. Measured from its base on the sea floor, Mauna Kea is taller than Mount Everest, the tallest mountain on land.

Cinder cone volcano

Cinder cone volcanoes are small volcanic cones made entirely of pyroclastic material from moderately explosive eruptions. The pyroclastic material forms steeper slopes with a narrower base than the lava flows of shield volcanoes, as you can see in this photo of the volcano Paricutín, in Mexico. Cinder cone volcanoes usually erupt for only a short time and often occur in clusters, commonly on the sides of shield and composite volcanoes. They erode quickly because the pyroclastic particles are not cemented together by lava.

Composite volcano

Composite volcanoes, sometimes referred to as *stratovolcanoes,* are one of the most common types of volcanoes. They form by explosive eruptions of pyroclastic material followed by quieter outpourings of lava. The combination of both types of eruptions forms alternating layers of pyroclastic material and lava. Composite volcanoes, such as Japan's Mount Fuji, shown here, have broad bases and sides that get steeper toward the summit.

Craters and Calderas

At the top of the central vent in most volcanoes is a funnel-shaped pit called a **crater.** (Craters are also the circular pits made by meteorite impacts.) The photograph of the cinder cone on the previous page shows a well-defined crater. A crater's funnel shape results from explosions of material out of the vent as well as the collapse of material from the crater's rim back into the vent. A **caldera** forms when a magma chamber that supplies material to a volcano empties and its roof collapses. This causes the ground to sink, leaving a large, circular depression, as shown in **Figure 6.**

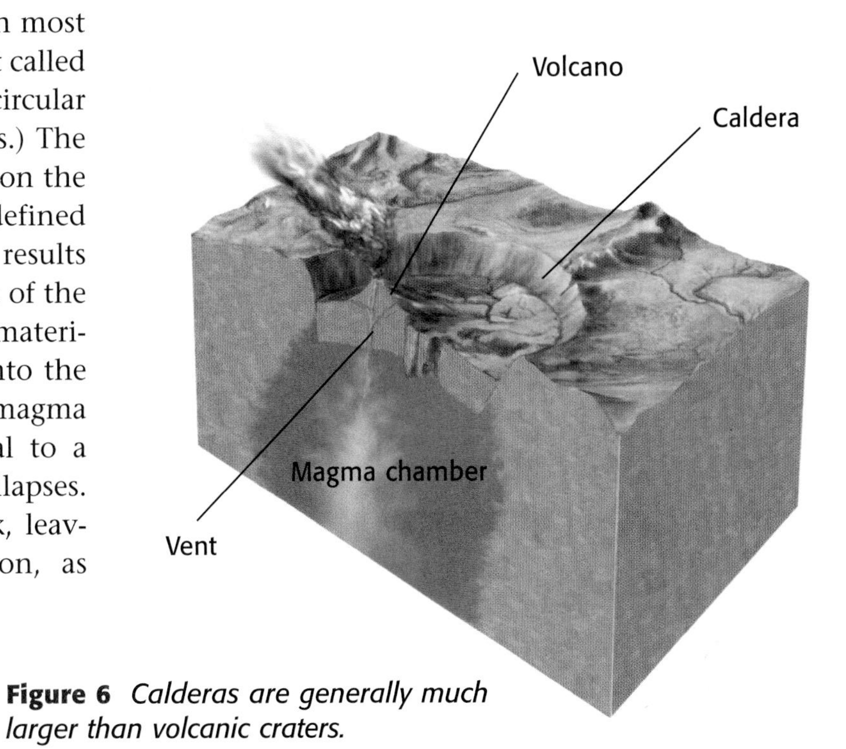

Figure 6 *Calderas are generally much larger than volcanic craters.*

Lava Plateaus

The most massive outpourings of lava do not come from individual volcanoes. Most of the lava on Earth's continents erupts from long cracks, or *fissures,* in the crust. In this non-explosive type of eruption, runny lava pours from a series of fissures and may spread evenly over thousands of square kilometers. The resulting landform is known as a *lava plateau.* The Columbia River Plateau, a lava plateau that formed about 15 million years ago, can be found in the northwestern United States.

SECTION REVIEW

1. Briefly explain why the ash from a volcanic eruption can be hazardous.
2. Why do cinder cone volcanoes have narrower bases and steeper sides than shield volcanoes?
3. **Comparing Concepts** Briefly describe the difference between a crater and a caldera.

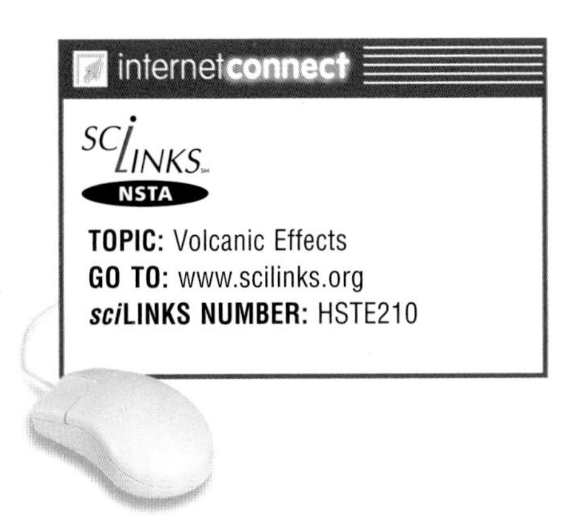

SECTION 3
READING WARM-UP

What Causes Volcanoes?

Terms to Learn

rift
hot spot

What You'll Do

- Describe the formation and movement of magma.
- Explain the relationship between volcanoes and plate tectonics.
- Summarize the methods scientists use to predict volcanic eruptions.

Scientists have learned a great deal over the years about what happens when a volcano erupts. Many of the results are dramatic and immediately visible. Unfortunately, understanding what causes a volcano to erupt in the first place is much more difficult. Scientists must rely on models based on rock samples and other data that provide insight into volcanic processes. Because it is so difficult to "see" what is going on deep inside the Earth, there are many uncertainties about why volcanoes form.

Reaction to Stress

1. Make a pliable "rock" by pouring 60 mL (1/4 cup) of **water** into a **plastic cup** and adding 150 mL of **cornstarch,** 15 mL (1 tbsp) at a time. Stir well after each addition.
2. Pour half of the cornstarch mixture into a **clear bowl.** Carefully observe how the "rock" flows. Be patient—this is a slow process!
3. Scrape the rest of the "rock" out of the cup with a **spoon.** Observe the behavior of the "rock" as you scrape.
4. What happened to the "rock" when you let it flow by itself? What happened when you put stress on the "rock"?
5. How is this pliable "rock" similar to the rock of the upper part of the mantle?

TRY at HOME

The Formation of Magma

You learned in the previous section that volcanoes form by the eruption of lava and pyroclastic material onto the Earth's surface. But the key to understanding why volcanoes erupt is understanding how magma forms. As you can see in **Figure 7,** volcanoes begin when magma collects in the deeper regions of the Earth's crust and in the uppermost layers of the mantle, the zone of intensely hot and pliable rock beneath the Earth's crust.

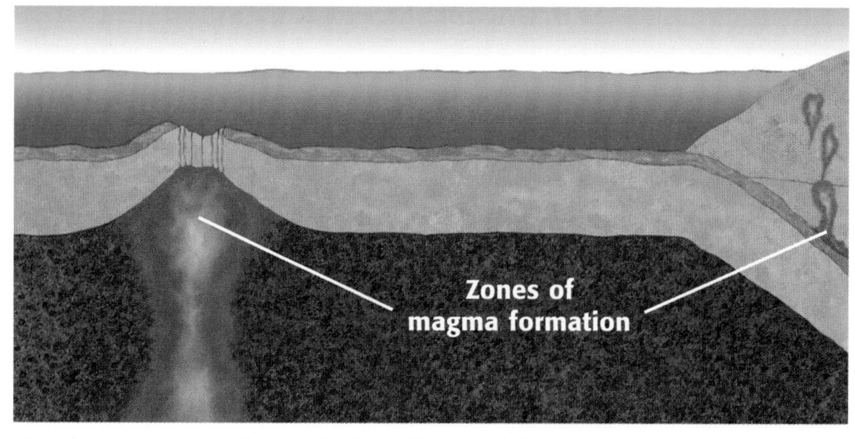

Figure 7 *Magma forms below the Earth's surface in a region that includes the lower crust and part of the upper mantle.*

Pressure and Temperature Although hot and pliable, the rock of the mantle is considered a solid. But the temperature of the mantle is high enough to melt almost any rock, so why doesn't it melt? The answer has to do with pressure. The weight of the rock above the mantle exerts a tremendous amount of pressure. This pressure keeps the atoms of mantle rock tightly packed, preventing the rock from changing into a liquid state. An increase in pressure raises the melting point of most materials.

As you can see in **Figure 8,** rock melts and forms magma when the temperature of the rock increases or when the pressure on the rock decreases. Because the temperature of the mantle is relatively constant, a decrease in pressure is usually what causes magma to form.

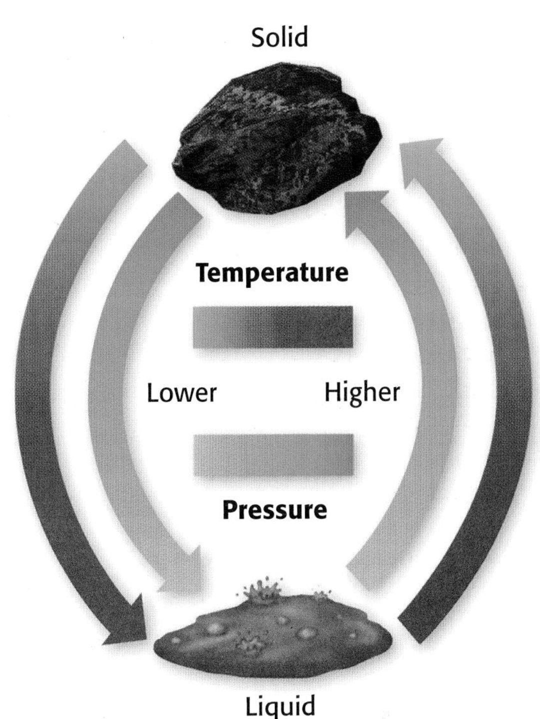

Figure 8 *This diagram shows how both pressure and temperature affect the formation of magma within the mantle.*

Density Once formed, the magma rises toward the surface of the Earth because it is less dense than the surrounding rock. Magma is commonly a mixture of liquid and solid mineral crystals and is therefore normally less dense than the completely solid rock that surrounds it. Like air bubbles that form on the bottom of a pan of boiling water, magma will rise toward the surface.

Self-Check

What two factors may cause solid rock to become magma? *(See page 216 to check your answer.)*

Where Volcanoes Form

The locations of volcanoes around the globe provide clues to how volcanoes form. The world map in **Figure 9** shows the location of the world's active volcanoes on land. It also shows tectonic plate boundaries. As you can see, a large number of the volcanoes lie directly on tectonic plate boundaries. In fact, the plate boundaries surrounding the Pacific Ocean have so many volcanoes that these boundaries together are called the *Ring of Fire.*

Why are most volcanoes on tectonic plate boundaries? These boundaries are where the plates either collide with one another or separate from one another. At these boundaries, it is easier for magma to travel upward through the crust. In other words, the boundaries are where the action is!

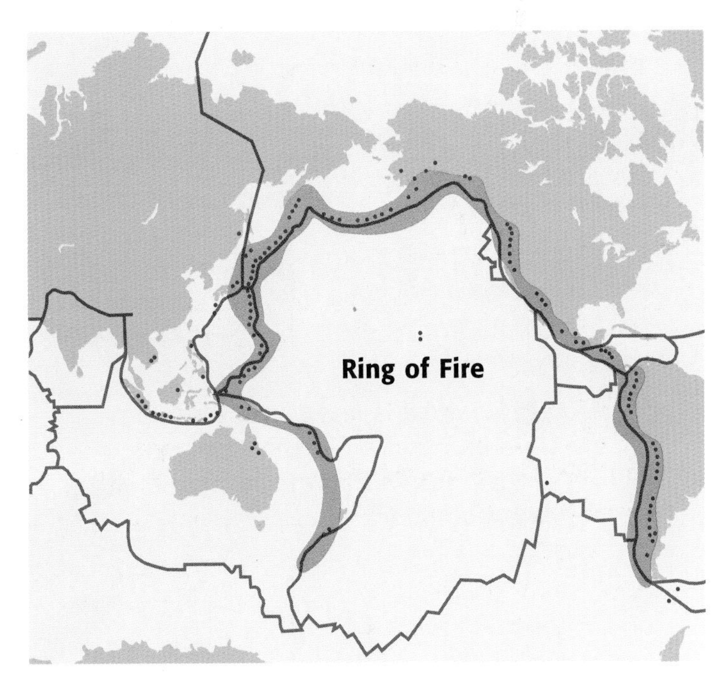

Figure 9 *Tectonic plate boundaries are likely places for volcanoes to form. The Ring of Fire contains nearly 75 percent of the world's active volcanoes on land.*

MATHBREAK

How Hot Is Hot?

Inside the Earth, magma can reach a burning-hot 1,400°C! You are probably more familiar with Fahrenheit temperatures, so convert 1,400°C to degrees Fahrenheit by using the formula below.

$$°F = \frac{9}{5}°C + 32$$

What is the magma's temperature in degrees Fahrenheit?

When Tectonic Plates Separate When two tectonic plates separate and move away from each other, a *divergent boundary* forms. As the tectonic plates separate, a deep crack, or **rift,** forms between the plates. Mantle material then rises to fill in the gap. Because the mantle material is now closer to the surface, the pressure on it decreases. This decrease in pressure causes the mantle rock to partially melt and become magma.

Because magma is less dense than the surrounding rock, it rises up through the rift. As the magma rises, it cools down, and the pressure on it decreases. So even though it becomes cooler as it rises, it remains molten because of the reduced pressure.

Magma continuously rises up through the rift between the separating plates and creates new crust. Although a few divergent boundaries exist on land, most are located on the ocean floor. There they produce long mountain chains called mid-ocean spreading centers, or *mid-ocean ridges.* **Figure 10** shows the process of forming such an underwater mountain range at a divergent boundary.

Figure 10 How Magma Forms at a Divergent Boundary

Mantle material rises to fill the space opened by separating tectonic plates. As the pressure decreases, the mantle begins to melt.

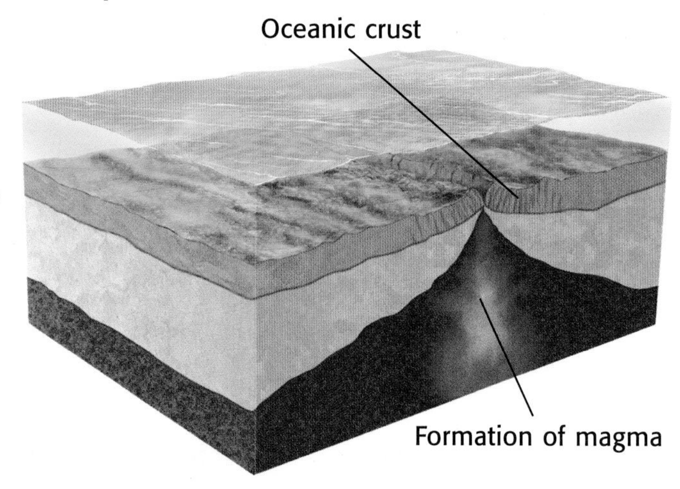

Because magma is less dense than the surrounding rock, it rises toward the surface, where it forms new crust on the ocean floor.

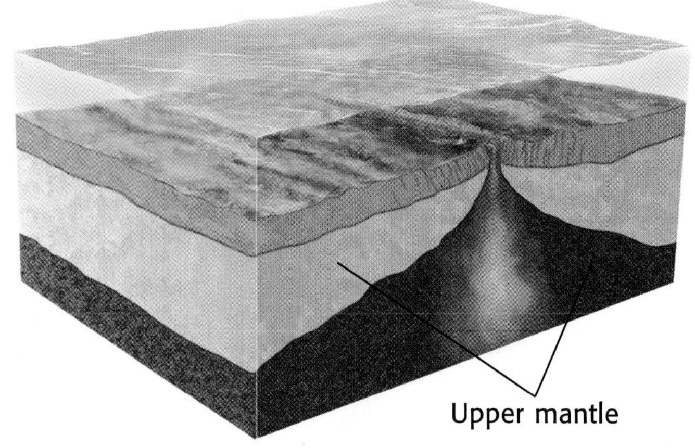

When Tectonic Plates Collide

If you slide two pieces of notebook paper into one another on a flat desktop, the papers will either buckle upward or one piece of paper will move under the other. This gives you an idea of what happens when tectonic plates collide. The place where two tectonic plates collide is called a *convergent boundary.*

Convergent boundaries are commonly located where oceanic plates collide with continental plates. The oceanic crust is denser and thinner and therefore moves underneath the continental crust. The movement of one tectonic plate under another is called *subduction,* shown in **Figure 11.**

As the descending oceanic crust scrapes past the continental crust, it sinks deeper into the mantle, getting hotter. As it does so, the pressure on the oceanic crust increases as well. The combination of increased heat and pressure causes the water contained in the oceanic crust to be released. The water then mixes with the mantle rock, which lowers the rock's melting point, causing it to melt.

Figure 11 How Magma Forms at a Convergent Boundary

1 *As the oceanic plate moves downward, some of the rock melts and forms magma.*

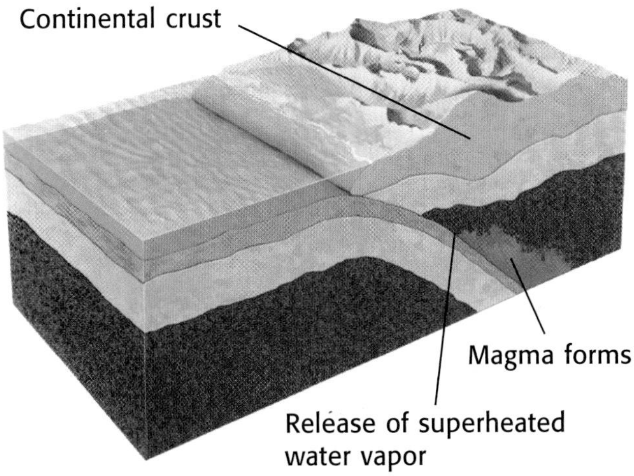

Continental crust
Magma forms
Release of superheated water vapor

2 *When magma is less dense than the surrounding rock, it rises toward the surface.*

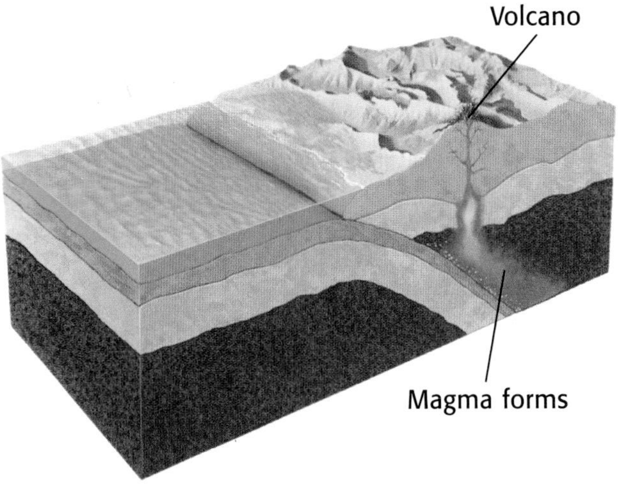

Volcano
Magma forms

Hot Spots

Not all magma develops along tectonic plate boundaries. For example, the Hawaiian Islands, some of the most well-known volcanoes on Earth, are nowhere near a plate boundary. The volcanoes of Hawaii and several other places on Earth are known as *hot spots.* **Hot spots** are places on the Earth's surface that are directly above columns of rising magma, called *mantle plumes.* Mantle plumes begin deep in the Earth, possibly at the boundary between the mantle and the core. Scientists are not sure what causes these plumes, but some think that a combination of heat conducted upward from the core and heat from radioactive elements keeps the plumes rising.

A hot spot often produces a long chain of volcanoes. This is because the mantle plume stays in the same spot, while the tectonic plate above moves over it. The Hawaiian Islands, for example, are riding on the Pacific plate, which is moving slowly to the northwest. **Figure 12** shows how a hot spot can form a chain of volcanic islands.

Figure 12 How a Hot Spot Forms Volcanoes

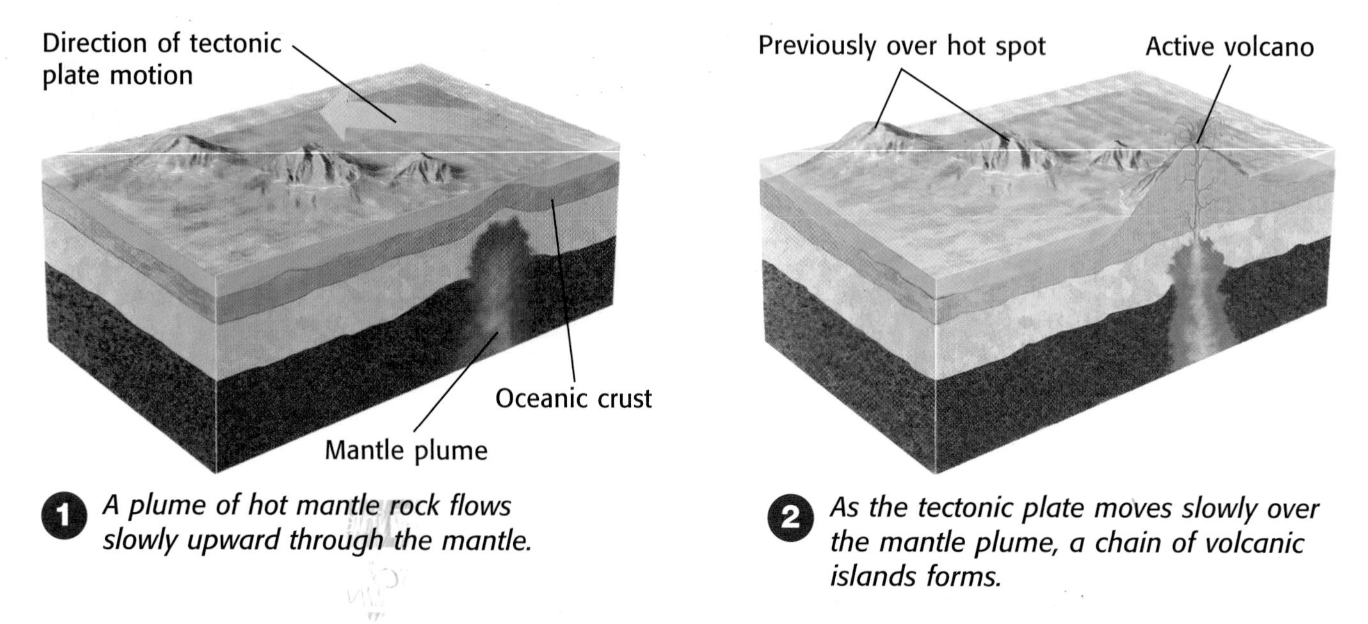

1 *A plume of hot mantle rock flows slowly upward through the mantle.*

2 *As the tectonic plate moves slowly over the mantle plume, a chain of volcanic islands forms.*

Predicting Volcanic Eruptions

To help predict volcanic eruptions, scientists classify volcanoes based on their eruption histories and on how likely it is that they will erupt again. *Extinct* volcanoes are those that have not erupted in recorded history and probably never will again. *Dormant* volcanoes are those that are not currently erupting but have erupted at some time in recorded history. *Active* volcanoes are those that are in the process of erupting or that show signs of erupting in the very near future.

Figure 13 *Seismographs help scientists determine when magma is moving beneath a volcano.*

Measuring Small Quakes Most active volcanoes produce small earthquakes as the magma within them moves upward and causes the surrounding rock to shift. Just before an eruption, the number and intensity of the small earthquakes increase, and the occurrence of quakes may be continuous. These earthquakes are measured with a *seismograph,* as shown in **Figure 13.**

Measuring Slope Measurements of a volcano's slope also give scientists clues with which to predict eruptions. For example, bulges in the volcano's slope may form as magma pushes against the inside of the volcano. By attaching an instrument called a *tiltmeter* to the surface of the volcano, scientists can detect small changes in the angle of the slope.

Measuring Volcanic Gases The outflow of volcanic gases from a volcano can also help scientists predict eruptions. Some scientists think that the ratio of certain gases, especially that of sulfur dioxide (SO_2) to carbon dioxide (CO_2), is important in predicting eruptions. They know that when this ratio changes, it is an indication that things are changing in the magma chamber down below! As you can see in **Figure 14,** collecting this type of data is often dangerous.

Figure 14 *As if getting this close to an active volcano is not dangerous enough, the gases that are being collected here are extremely poisonous.*

Measuring Temperature from Orbit Some of the newest methods scientists are using to predict volcanic eruptions rely on satellite images. Many of these images record infrared radiation, which allows scientists to measure changes in temperature over time. They are taken from satellites orbiting more than 700 km above the Earth. By analyzing images taken at different times, scientists can determine if the site is getting hotter as magma pushes closer to the surface.

SECTION REVIEW

1. How does pressure determine whether the mantle is solid or liquid?
2. Describe a technology scientists use to predict volcanic eruptions.
3. **Interpreting Illustrations** Figure 9, shown earlier in this chapter, shows the locations of active volcanoes on land. Describe where on the map you would plot the location of underwater volcanoes and why. (Do not write in this book.)

Calling an Evacuation?

Although scientists have learned a lot about volcanoes, they cannot predict eruptions with total accuracy. Sometimes there are warning signs before an eruption, but often there are none. Imagine that you are the mayor of a town near a large volcano, and a geologist warns you that an eruption is probable. You realize that ordering an evacuation of your town could be an expensive embarrassment if the volcano doesn't erupt. But if you decide to keep quiet, people could be in serious danger if the volcano does erupt. Considering the social and economic consequences of your decision, your job is perhaps even more difficult. What would you do?

Discovery Lab

Some Go "Pop," Some Do Not

Volcanic eruptions range from mild to violent. When volcanoes erupt, the rocks left behind provide information to scientists studying Earth's crust. Mild, or nonexplosive, eruptions produce lava that is low in silica and has a low viscosity. In nonexplosive eruptions, lava simply flows down the side of the volcano. Explosive eruptions, on the other hand, do not produce much lava. Instead, the explosions hurl ash and debris into the air. The rocks left behind are light in color, are high in silica, and have a high viscosity. These rocks help geologists find out what the crust below the volcanoes is made of.

MATERIALS

- graph paper
- metric ruler
- red, yellow, and orange colored pencils or markers

Procedure

1. Copy the map on the next page onto graph paper or recreate it using a computer. Be sure to line the grid up properly.
2. Locate each volcano from the list by drawing a circle with a diameter of about 1 cm in the proper location on your copy of the map. Use the latitude and longitude grids to help you.
3. Review all the eruptions for each volcano. For each explosive eruption, color the circle red. For each quiet volcano, color the circle yellow. For volcanoes that have erupted in both ways, color the circle orange.

Analysis

4. According to your map, where are volcanoes that always have nonexplosive eruptions located?
5. Where are volcanoes that always erupt explosively located? Where are volcanoes that erupt in both ways located?
6. If volcanoes get their magma from the crust below them, what can you infer about the silica content of Earth's crust under the oceans?
7. What is the composition of the crust under the continents? How do we know?
8. What is the source of rocks for volcanoes that erupt in both ways? How do you know?
9. Do the locations of volcanoes that erupt in both ways make sense based on your answers to items 7 and 8? Explain.

Volcanic Activity Chart		
Volcano name	**Location**	**Description**
Mount St. Helens	46°N 122°W	An explosive eruption blew the top off the mountain. Light-colored ash covered thousands of square kilometers. Another eruption sent a lava flow down the southeast side of the mountain.
Kilauea	19°N 155°W	One small eruption sent a lava flow along 12 km of highway.
Rabaul caldera	4°S 152°E	Explosive eruptions have caused tsunamis and have left 1–2 m of ash on nearby buildings.
Popocatépetl	19°N 98°W	During one explosion, Mexico City closed the airport for 14 hours because huge columns of ash made it too difficult for pilots to see. Eruptions from this volcano have also caused damaging avalanches.
Soufriere Hills	16°N 62°W	Small eruptions have sent lava flows down the hills. Other explosive eruptions have sent large columns of ash into the air.
Long Valley caldera	37°N 119°W	Explosive eruptions have sent ash into the air.
Okmok	53°N 168°W	Recently, there have been slow lava flows from this volcano. Twenty-five hundred years ago, ash and debris exploded from the top of this volcano.
Pavlof	55°N 161°W	Eruption clouds have been sent 200 m above the summit. Eruptions have sent ash columns 10 km into the air. Occasionally, small eruptions have caused lava flows.
Fernandina	42°N 12°E	Eruptions have ejected large blocks of rock from this volcano.
Mount Pinatubo	15°N 120°E	Ash and debris from an explosive eruption destroyed homes, crops, and roads within 52,000 km^2 around the volcano.

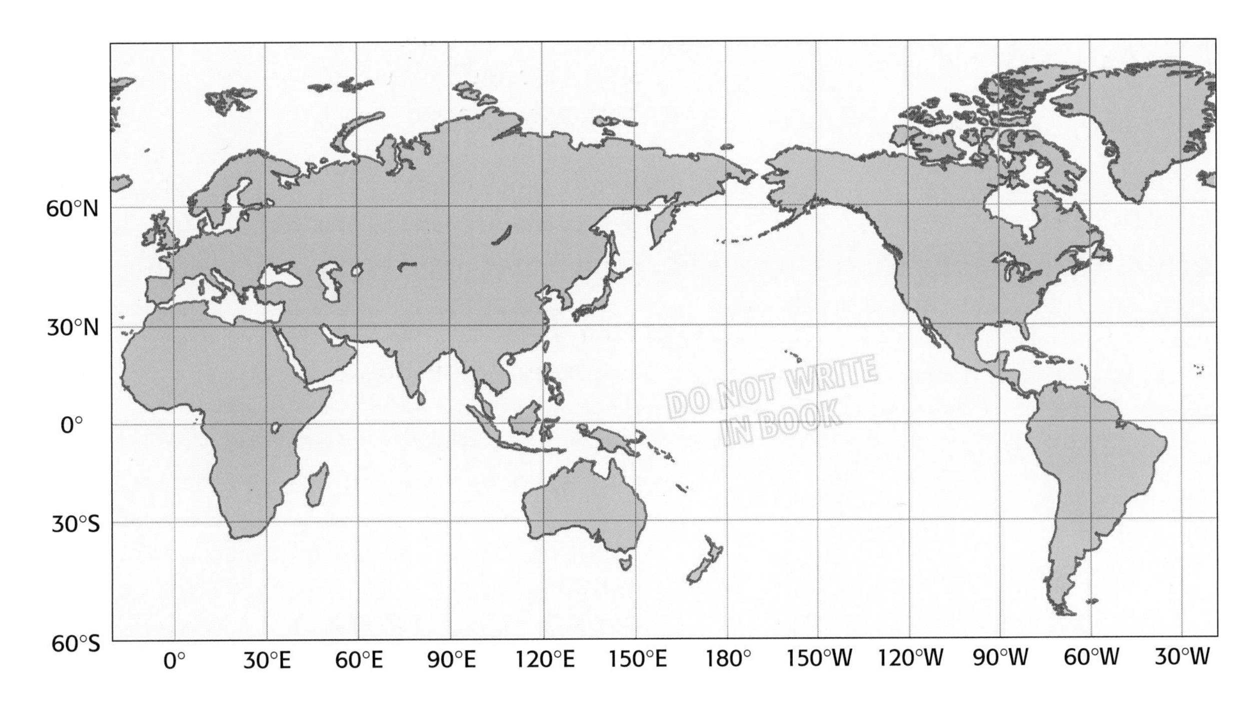

Chapter Highlights

SECTION 1

Vocabulary

volcano *(p. 148)*

lava *(p. 148)*

pyroclastic material *(p. 151)*

Section Notes

- Volcanoes erupt both explosively and nonexplosively.
- The characteristics of a volcanic eruption are largely determined by the type of magma within the volcano.
- The amount of silica in magma determines whether it is thin and fluid or thick and stiff.
- Lava hardens into characteristic features that range from smooth to jagged, depending on how thick the lava is and how quickly it flows.
- Pyroclastic material, or volcanic debris, consists of solid pieces of the volcano as well as magma that solidifies as it travels through the air.

SECTION 2

Vocabulary

shield volcano *(p. 154)*

cinder cone volcano *(p. 154)*

composite volcano *(p. 154)*

crater *(p. 155)*

caldera *(p. 155)*

Section Notes

- The effects of volcanic eruptions are felt both locally and around the world.
- Volcanic mountains can be classified according to their composition and overall shape.
- Craters are funnel-shaped pits that form around the central vent of a volcano. Calderas are large bowl-shaped depressions formed by a collapsed magma chamber.

Skills Check

Math Concepts

CONVERTING TEMPERATURE SCALES So-called low-temperature magmas can be 1,100°C. Just how hot is such a magma? If you are used to measuring temperature in degrees Fahrenheit, you can use a simple formula to find out.

$$°F = \frac{9}{5}°C + 32$$

$$°F = \frac{9}{5}(1{,}100) + 32$$

$$°F = 1{,}980 + 32 = 2{,}012$$

$$2{,}012°F = 1{,}100°C$$

Visual Understanding

CALDERAS Calderas are caused by the release of massive amounts of magma from beneath the Earth's surface. When the volume of magma decreases, it no longer exerts pressure to hold the ground up. As a result, the ground sinks, forming a caldera.

SECTION 2

- In the largest type of volcanic eruption, lava simply pours from long fissures in the Earth's crust to form lava plateaus.

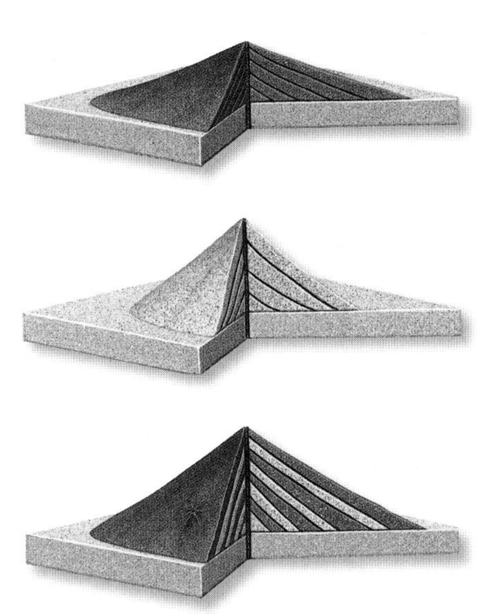

SECTION 3

Vocabulary

rift *(p. 158)*

hot spot *(p. 159)*

Section Notes

- Volcanoes result from magma formed in the mantle.
- When pressure is reduced, some of the solid rock of the already hot mantle melts to form magma.
- Because it is less dense than the surrounding rock, magma rises to the Earth's surface. It either erupts as lava or solidifies in the crust.
- Most volcanic activity takes place along tectonic plate boundaries, where plates either separate or collide.
- Volcanoes also occur at hot spots. Chains of volcanic islands can form when tectonic plates move relative to the hot spot.
- Volcanic eruptions cannot be predicted with complete accuracy. But scientists now have several methods of forecasting future eruptions.

Labs

Volcano Verdict *(p. 186)*

internet**connect**

GO TO: go.hrw.com

Visit the **HRW** Web site for a variety of learning tools related to this chapter. Just type in the keyword:

KEYWORD: HSTVOL

GO TO: www.scilinks.org

Visit the **National Science Teachers Association** on-line Web site for Internet resources related to this chapter. Just type in the *sci*LINKS number for more information about the topic:

TOPIC: Volcanic Eruptions — ***sci*LINKS NUMBER:** HSTE205

TOPIC: Volcanic Effects — ***sci*LINKS NUMBER:** HSTE210

TOPIC: What Causes Volcanoes? — ***sci*LINKS NUMBER:** HSTE215

TOPIC: The Ring of Fire — ***sci*LINKS NUMBER:** HSTE220

Chapter Review

USING VOCABULARY

For each pair of terms listed below, explain the difference in their meanings.

1. caldera/crater
2. lava/magma
3. lava/pyroclastic material
4. vent/rift
5. cinder cone volcano/shield volcano

UNDERSTANDING CONCEPTS

Multiple Choice

6. The type of magma that often produces a violent eruption can be described as
   a. thin due to high silica content.
   b. thick due to high silica content.
   c. thin due to low silica content.
   d. thick due to low silica content.

7. When lava hardens quickly to form ropy formations, it is called
   a. aa lava.
   b. pahoehoe lava.
   c. pillow lava.
   d. blocky lava.

8. Volcanic dust and ash can remain in the atmosphere for months or years, causing
   a. decreased solar reflection and higher temperatures.
   b. increased solar reflection and lower temperatures.
   c. decreased solar reflection and lower temperatures.
   d. increased solar reflection and higher temperatures.

9. Mount St. Helens, in Washington, covered the city of Spokane with tons of ash. Its eruption would most likely be described as
   a. nonexplosive, producing lava.
   b. explosive, producing lava.
   c. nonexplosive, producing pyroclastic material.
   d. explosive, producing pyroclastic material.

10. Magma forms within the mantle most often as a result of
    a. high temperature and high pressure.
    b. high temperature and low pressure.
    c. low temperature and high pressure.
    d. low temperature and low pressure.

11. At divergent plate boundaries,
    a. heat from the Earth's core produces mantle plumes.
    b. oceanic plates sink, causing magma to form.
    c. tectonic plates move apart.
    d. hot spots produce volcanoes.

12. A theory that helps to explain the causes of both earthquakes and volcanoes is the theory of
    a. pyroclastics.
    b. plate tectonics.
    c. climatic fluctuation.
    d. mantle plumes.

Short Answer

13. Briefly describe two methods that scientists use to predict volcanic eruptions.

14. Describe how differences in magma affect volcanic eruptions.

15. Along what types of tectonic plate boundaries are volcanoes generally found? Why?

16. Describe the characteristics of the three types of volcanic mountains.

Concept Mapping

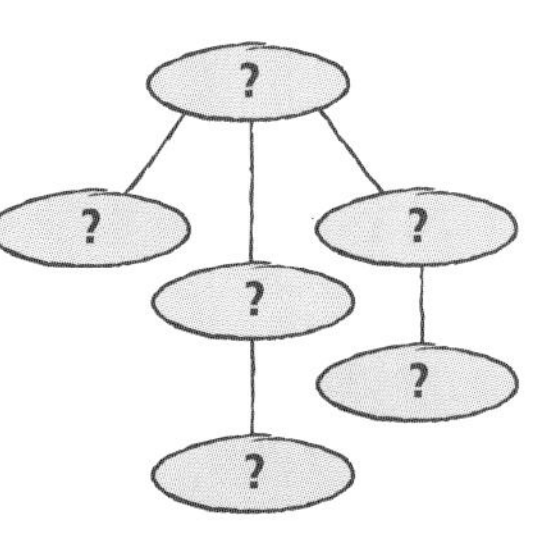

17. Use any of the terms from the vocabulary lists in Chapter Highlights to construct a concept map that illustrates the relationship between types of magma, the eruptions they produce, and the shapes of the volcanoes that result.

CRITICAL THINKING AND PROBLEM SOLVING

Write one or two sentences to answer the following questions:

18. Imagine that you are exploring a volcano that has been dormant for some time. You begin to keep notes on the types of volcanic debris you encounter as you walk. Your first notes describe volcanic ash, and later your notes describe lapilli. In what direction would you most likely be traveling—toward or away from the crater? Explain.

19. Loihi is a future Hawaiian island in the process of forming on the ocean floor. Considering how this island chain formed, tell where you think the new volcanic island will be located and why.

20. What do you think would happen to the Earth's climate if volcanic activity increased to 10 times its current level?

MATH IN SCIENCE

21. Midway Island is 1,935 km northwest of Hawaii. If the Pacific plate is moving to the northwest at 9 cm/yr, how long ago was Midway Island located over the hot spot that formed it?

INTERPRETING GRAPHICS

The following graph illustrates the average change in temperature above or below normal for a community over several years.

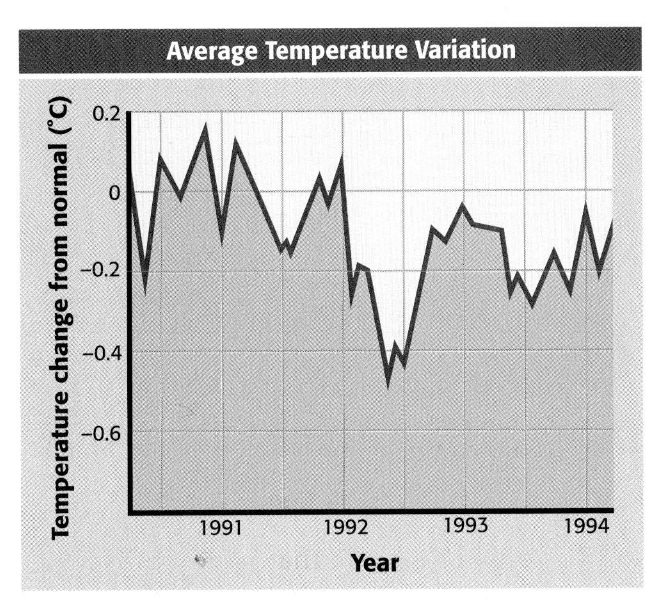

22. If the variation in temperature over the years was influenced by a major volcanic eruption, when did the eruption most likely take place? Explain.

23. If the temperature were plotted only in yearly intervals rather than several times per year, how might your interpretation be different?

Take a minute to review your answers to the Pre-Reading Questions found at the bottom of page 146. Have your answers changed? If necessary, revise your answers based on what you have learned since you began this chapter.

Science, Technology, and Society

Robot in the Hot Seat

Scientists have to be calm, cool, and collected to study active volcanoes. But the recently cooled magma in a volcanic crater isn't the most hospitable location for scientific study. What kind of daredevil would run the risk of creeping along a crater floor? A volcanologist like *Dante II,* that's who!

Hot Stuff

A volcano crater may seem empty after a volcano erupts, but it is in no way devoid of volcanic information. Gases hissing up through the crater floor give scientists clues about the molten rock underneath, which may help them understand how and why volcanoes erupt repeatedly. But these gases may be poisonous or scalding hot, and the crater's floor can crack or shift at any time. Over the years, dozens of scientists have been seriously injured or killed while trying to explore volcano craters. Obviously, volcanologists needed some help studying the steamy abyss.

▲ *Dante II*

Getting a Robot to Take the Heat

Enter *Dante II,* an eight-legged robot with cameras for eyes and computers for a brain. In 1994, led by a team of scientists from NASA, Carnegie Mellon University, and the Alaskan Volcano Observatory, *Dante II* embarked on its first mission. It climbed into a breach called Crater Peak on the side of Mount Spurr, an active volcano in Alaska. Anchored at the crater's rim by a strong cable, *Dante II* was controlled partly by internal computers and partly by a team of scientists. The team communicated with the robot through a satellite link and Internet connections. *Dante II* moved very slowly, taking pictures and collecting scientific data. It was equipped with gas sensors that provided continuous readings of the crater gases. It performed the tasks human scientists could not, letting the humans keep their cool.

Mission Accomplished?

During its expedition, *Dante II* encountered large rocks, some of which were as big as the robot itself. In addition, while climbing out of the volcano, *Dante II* slipped and fell, damaging one of its legs. Eventually *Dante II* had to be rescued by helicopter because its support cable broke. Despite these obstacles, *Dante II* was able to gather valuable data from the volcano's crater.

Dante II's mission also met one of NASA's objectives: to prove that robots could be used successfully to explore extreme terrain, such as that found on planetary surfaces. *Dante II* paved the way for later robotic projects, such as the exploration of the surface of Mars by the *Sojourner* rover in 1997.

Write About It

▶ Write a proposal for a project in which a robot is used to explore a dangerous place. Don't forget to include what types of data the robot would be collecting.

EARTH SCIENCE • LIFE SCIENCE

Europa: Life on a Moon?

Smooth and brownish white, one of Jupiter's moons, Europa, has fascinated scientists and science-fiction writers for decades. More recently, scientists were excited by tantalizing images from the Galileo Europa Mission. Could it be that life is lurking (or sloshing) beneath Europa's surface?

An Active History

Slightly smaller than Earth's moon, Europa is the fourth largest of Jupiter's moons. It is unusual among other bodies in the solar system because of its extraordinarily smooth surface. The ridges and brownish channels that crisscross Europa's smooth surface may tell a unique story—the surface appears to be a slushy combination of ice and water. Some scientists think that the icy ridges and channels are ice floes left over from ancient volcanoes that erupted water! The water flowed over Europa's surface and froze, like lava flows and cools on Earth's surface.

A Slushy Situation

Scientists speculate that Europa's surface consists of thin tectonic plates of ice floating on a layer of slush or water. These plates, which would look like icy rafts floating in an ocean of slush, have been compared to giant glaciers floating in polar regions on Earth.

Where plates push together, the material of the plates may crumple, forming an icy ridge. Where plates pull apart, warmer liquid mixed with darker silicates may erupt toward the surface and freeze, forming the brownish icy channels that create Europa's cracked cue-ball appearance.

◀ *Europa looks like a cracked cue ball.*

Life on Europa?

These discoveries have led scientists to consider an exciting possibility: Does Europa have an environment that could support primitive life-forms? In general, at least three things are necessary for life as we know it to develop—water, organic compounds (substances that contain carbon), and heat. Europa has water, and organic compounds are fairly common in the solar system. But is it hot enough? Europa's slushy nature suggests a warm interior. One theory is that the warmth is the result of Jupiter's strong gravitational pull on Europa. Another theory is that warmth is brought to Europa's surface by convection heating.

So does Europa truly satisfy the three requirements for life? The answer is still unknown, but the sloshing beneath Europa's surface has sure heightened some scientists' curiosity!

If You Were in Charge . . .

▶ If you were in charge of NASA's space-exploration program, would you send a spacecraft to look for life on Europa? (Remember that this would cost millions of dollars and would mean sacrificing other important projects!) Explain your answer.

SAFETY FIRST!

Exploring, inventing, and investigating are essential to the study of science. However, these activities can also be dangerous. To make sure that your experiments and explorations are safe, you must be aware of a variety of safety guidelines.

You have probably heard of the saying, "It is better to be safe than sorry." This is particularly true in a science classroom where experiments and explorations are being performed. Being uninformed and careless can result in serious injuries. Don't take chances with your own safety or with anyone else's.

Following are important guidelines for staying safe in the science classroom. Your teacher may also have safety guidelines and tips that are specific to your classroom and laboratory. Take the time to be safe.

Safety Rules!

Start Out Right

Always get your teacher's permission before attempting any laboratory exploration. Read the procedures carefully, and pay particular attention to safety information and caution statements. If you are unsure about what a safety symbol means, look it up or ask your teacher. You cannot be too careful when it comes to safety. If an accident does occur, inform your teacher immediately, regardless of how minor you think the accident is.

If you are instructed to note the odor of a substance, wave the fumes toward your nose with your hand. Never put your nose close to the source.

Safety Symbols

All of the experiments and investigations in this book and their related worksheets include important safety symbols to alert you to particular safety concerns. Become familiar with these symbols so that when you see them, you will know what they mean and what to do. It is important that you read this entire safety section to learn about specific dangers in the laboratory.

Eye Safety

Wear safety goggles when working around chemicals, acids, bases, or any type of flame or heating device. Wear safety goggles any time there is even the slightest chance that harm could come to your eyes. If any substance gets into your eyes, notify your teacher immediately, and flush your eyes with running water for at least 15 minutes. Treat any unknown chemical as if it were a dangerous chemical. Never look directly into the sun. Doing so could cause permanent blindness.

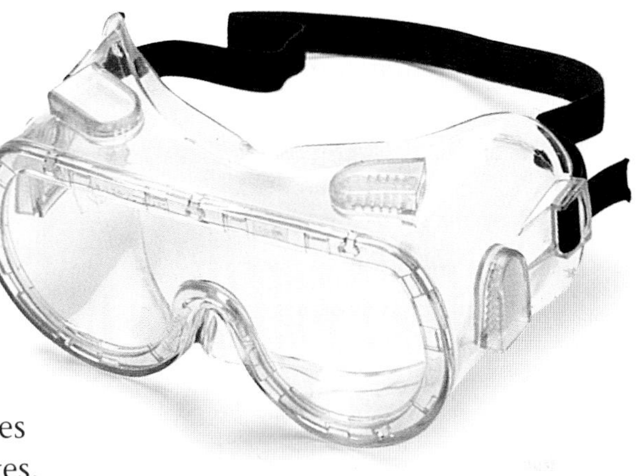

Avoid wearing contact lenses in a laboratory situation. Even if you are wearing safety goggles, chemicals can get between the contact lenses and your eyes. If your doctor requires that you wear contact lenses instead of glasses, wear eye-cup safety goggles in the lab.

Safety Equipment

Know the locations of the nearest fire alarms and any other safety equipment, such as fire blankets and eyewash fountains, as identified by your teacher, and know the procedures for using them.

Be extra careful when using any glassware. When adding a heavy object to a graduated cylinder, tilt the cylinder so the object slides slowly to the bottom.

Neatness

Keep your work area free of all unnecessary books and papers. Tie back long hair, and secure loose sleeves or other loose articles of clothing, such as ties and bows. Remove dangling jewelry. Don't wear open-toed shoes or sandals in the laboratory. Never eat, drink, or apply cosmetics in a laboratory setting. Food, drink, and cosmetics can easily become contaminated with dangerous materials.

Certain hair products (such as aerosol hair spray) are flammable and should not be worn while working near an open flame. Avoid wearing hair spray or hair gel on lab days.

Sharp/Pointed Objects

Use knives and other sharp instruments with extreme care. Never cut objects while holding them in your hands. Place objects on a suitable work surface for cutting.

Heat

Wear safety goggles when using a heating device or a flame. Whenever possible, use an electric hot plate as a heat source instead of an open flame. When heating materials in a test tube, always angle the test tube away from yourself and others. In order to avoid burns, wear heat-resistant gloves whenever instructed to do so.

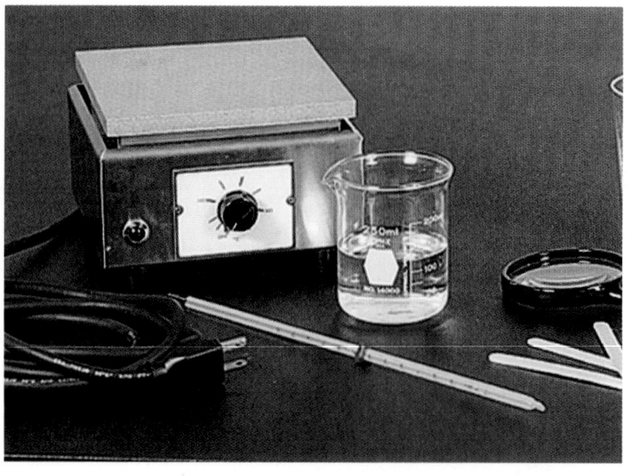

Chemicals

Wear safety goggles when handling any potentially dangerous chemicals, acids, or bases. If a chemical is unknown, handle it as you would a dangerous chemical. Wear an apron and safety gloves when working with acids or bases or whenever you are told to do so. If a spill gets on your skin or clothing, rinse it off immediately with water for at least 5 minutes while calling to your teacher.

Never mix chemicals unless your teacher tells you to do so. Never taste, touch, or smell chemicals unless you are specifically directed to do so. Before working with a flammable liquid or gas, check for the presence of any source of flame, spark, or heat.

Electricity

Be careful with electrical cords. When using a microscope with a lamp, do not place the cord where it could trip someone. Do not let cords hang over a table edge in a way that could cause equipment to fall if the cord is accidentally pulled. Do not use equipment with damaged cords. Be sure your hands are dry and that the electrical equipment is in the "off" position before plugging it in. Turn off and unplug electrical equipment when you are finished.

Animal Safety

Always obtain your teacher's permission before bringing any animal into the school building. Handle animals only as your teacher directs. Always treat animals carefully and with respect. Wash your hands thoroughly after handling any animal.

Plant Safety

Do not eat any part of a plant or plant seed used in the laboratory. Wash hands thoroughly after handling any part of a plant. When in nature, do not pick any wild plants unless your teacher instructs you to do so.

Glassware

Examine all glassware before use. Be sure that glassware is clean and free of chips and cracks. Report damaged glassware to your teacher. Glass containers used for heating should be made of heat-resistant glass.

Using the Scientific Method

Geologists often use a technique called *core sampling* to learn what underground rock layers look like. This technique involves drilling several holes in the ground in different places and taking samples of the underground rock or soil. Geologists then compare the samples from each hole to construct a diagram that shows the bigger picture.

In this activity, you will model the process geologists use to diagram underground rock layers. You will first use modeling clay to form a rock-layer model. You will then exchange models with a classmate, take core samples, and draw a diagram of your classmate's layers.

Materials

- 3 colored pencils or markers
- nontransparent pan or box
- modeling clay in three colors
- 1/2 in. PVC pipe
- plastic knife

Ask a Question

1. Can unseen features be revealed by sampling parts of the whole?

Form a Hypothesis

2. Form a hypotheses on whether taking core samples from several locations will give a good indication of the entire hidden feature.

Test the Hypothesis

3. To test your hypothesis, you will take core samples from a model of underground rock layers, draw a diagram of the entire rock-layer sequence, and then compare your drawing with the actual model.

Build a Model

The model rock layers should be formed out of view of the classmates who will be taking the core samples.

4. Form a plan for your rock layers, and sketch the layers in your ScienceLog. Your sketch should include the three colors in several layers of varying thicknesses.

5. In the pan or box, mold the clay into the shape of the lowest layer in your sketch.
6. Repeat step 5 for each additional layer of clay. You now have a rock-layer model. Exchange models with a classmate.

Collect Data

7. Choose three places on the surface of the clay to drill holes. The holes should be far apart and in a straight line. (You do not need to remove the clay from the pan or box.)
8. Use the PVC pipe to "drill" a vertical hole in the clay at one of the chosen locations by slowly pushing the pipe through all the layers of clay. Slowly remove the pipe.
9. Remove the core sample from the pipe by gently pushing the clay out of the pipe with an unsharpened pencil.
10. Draw the core sample in your ScienceLog, and record your observations. Be sure to use a different color of pencil or marker for each layer.
11. Repeat steps 8–10 for the next two core samples. Make sure your drawings are side by side in your ScienceLog in the same order as the samples in the model.

Analyze the Results

12. Look at the pattern of rock layers in each of your core samples. Think about how the rock layers between the core samples might look. Then construct a diagram of the rock layers.
13. Complete your diagram by coloring the rest of the rock layers.

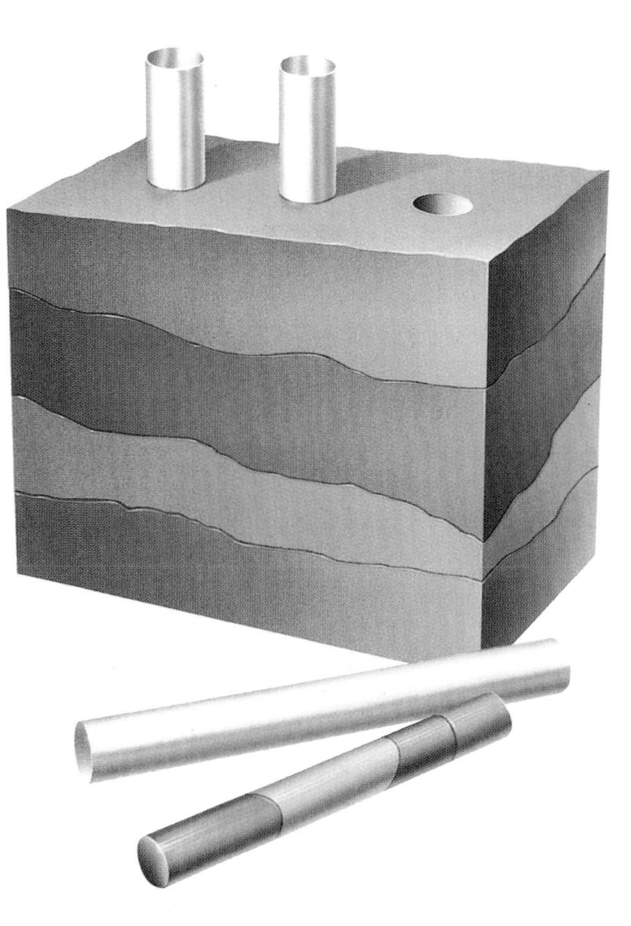

Draw Conclusions

14. Use the plastic knife to cut the clay model along a line connecting the three holes and remove one side of the model. The rock layers should now be visible.
15. How well does your rock-layer diagram match the model? Explain.
16. Is it necessary to revise your diagram from step 13? If so, how?
17. Do your conclusions support your hypothesis? Why or why not?

Going Further

What are two ways that the core-sampling method could be improved?

Using Scientific Methods

Is It Fool's Gold?—A Dense Situation

Have you heard of fool's gold? Maybe you've seen a piece of it. This notorious mineral was often passed off as real gold. There are, however, simple tests you can do to keep from being tricked. Minerals can be identified by their properties. Some properties, such as color, vary between different samples of the same mineral. Other properties, such as density and specific gravity, remain consistent from one sample to another. In this activity, you will try to verify the identity of some mineral samples.

Materials

- spring scale
- ring stand
- pyrite sample
- galena sample
- balance
- string
- 400 mL beaker
- 400 mL of water

Ask a Question

1. How can I determine if an unknown mineral is not gold or silver?

Make Observations

2. Copy the data table below into your ScienceLog. Use it to record your observations.

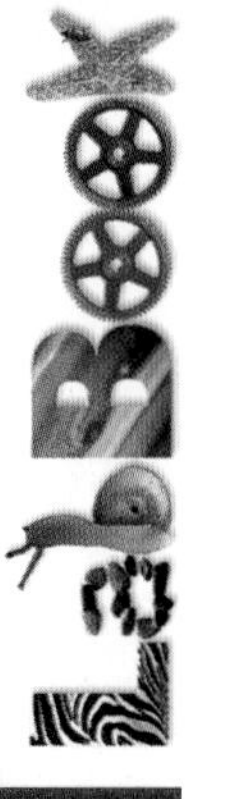

Observation Chart

Measurement	Galena	Pyrite
Mass in air (g)		
Weight in air (N)		
Beginning volume of water (mL)		
Final volume of water (mL)		
Volume of mineral (mL)		
Weight in water (N)		

3. Find the mass of each sample by laying the mineral on the balance. Record the mass of each in your data table.
4. Attach the spring scale to the ring stand.
5. Tie a string around the sample of galena, leaving a loop at the loose end. Suspend the galena from the spring scale, and find its weight in air. Do not remove the sample from the spring scale yet. Enter these data in your data table.

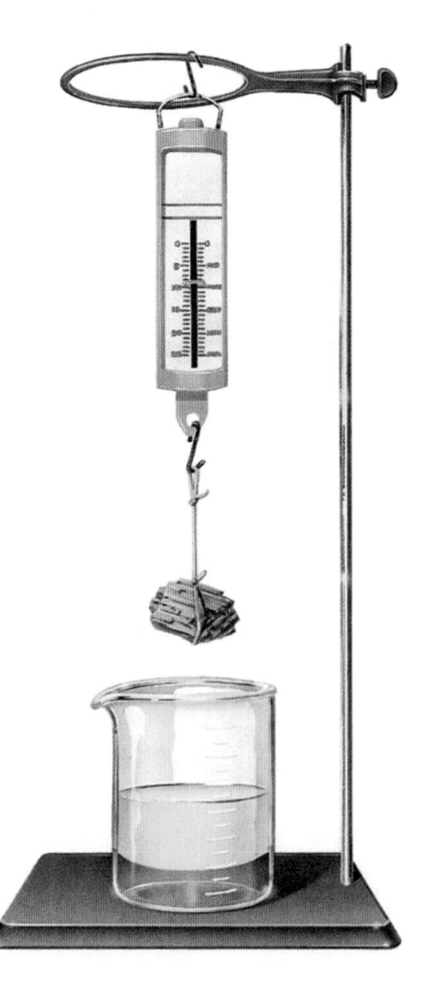

6. Fill a beaker halfway with water. Record the beginning volume of water in your data table.
7. Carefully lift the beaker around the galena until the mineral is completely submerged. Be careful not to splash any water out of the beaker! Be sure the mineral does not touch the beaker.
8. Record the new volume and weight in your data table.
9. Subtract the original volume of water from the new volume to find the amount of water displaced by the mineral. This is the volume of the mineral sample itself. Record this value in your data table.
10. Repeat steps 5–9 for the sample of pyrite.

Analyze the Results

11. Copy the data table below into your ScienceLog. **Note:** 1 mL = 1 cm^3
12. Use the following equations to calculate the density and specific gravity of each mineral, and record your answers in your data table.

$$\text{Density} = \frac{\text{mass in air}}{\text{volume}}$$

$$\text{Specific gravity} = \frac{\text{weight in air}}{\text{weight in air} - \text{weight in water}}$$

Mineral	Density (g/cm^3)	Specific gravity
Silver	10.5	10.5
Galena		
Pyrite		
Gold	19.3	19.3

Draw Conclusions

13. The density of pure gold is 19.3 g/cm^3. How can you use this information to prove that your sample of pyrite is not gold?
14. The density of pure silver is 10.5 g/cm^3. How can you use this information to prove that your sample of galena is not silver?
15. If you found a gold-colored nugget, how could you find out if the nugget was real gold or fool's gold?

Crystal Growth

Magma forms deep below the Earth's surface at depths of 25 to 160 km and at extremely high temperatures. Some magma reaches the surface and cools quickly. Other magma gets trapped in cracks or magma chambers beneath the surface and cools very slowly. When magma cools slowly, large, well-developed crystals form. On the other hand, when magma erupts onto the surface, thermal energy is lost rapidly to the air or water. There is not enough time for large crystals to grow. The size of the crystals found in igneous rocks gives geologists clues about where and how the crystals formed.

In this experiment, you will demonstrate how the rate of cooling affects the size of crystals in igneous rocks by cooling crystals of magnesium sulfate at two different rates.

Materials

- heat-resistant gloves
- 400 mL beaker
- 200 mL of tap water
- hot plate
- Celsius thermometer
- magnesium sulfate ($MgSO_4$) (Epsom salts)
- magnifying lens
- pointed laboratory scoop
- medium test tube
- distilled water
- watch or clock
- aluminum foil
- test-tube tongs
- dark marker
- masking tape
- basalt
- pumice
- granite

Make a Prediction

1. Suppose you have two solutions that are identical in every way except for temperature. How will the temperature of a solution affect the size of the crystals and the rate at which they form?

Make Observations

2. Put on your gloves, apron, and goggles.
3. Fill the beaker halfway with tap water. Place the beaker on the hot plate, and let it begin to warm. The temperature of the water should be between 40°C and 50°C.
   Caution: Make sure the hot plate is away from the edge of the lab table.
4. Examine two or three crystals of the magnesium sulfate with your magnifying lens. In your ScienceLog, describe the color, shape, luster, and other interesting features of the crystals.
5. Draw a sketch of the magnesium sulfate crystals in your ScienceLog.

Conduct an Experiment

6. Use the pointed laboratory scoop to fill the test tube about halfway with the magnesium sulfate. Add an equal amount of distilled water.

7. Hold the test tube in one hand, and use one finger from your other hand to tap the test tube gently. Observe the solution mixing as you continue to tap the test tube.

8. Place the test tube in the beaker of hot water, and heat it for approximately 3 minutes.
   Caution: Be sure to direct the opening of the test tube away from you and other students.

9. While the test tube is heating, shape your aluminum foil into two small boatlike containers by doubling the foil and turning up each edge.

10. If all the magnesium sulfate is not dissolved after 3 minutes, tap the test tube again, and heat it for 3 more minutes.
    Caution: Use the test-tube tongs to handle the hot test tube.

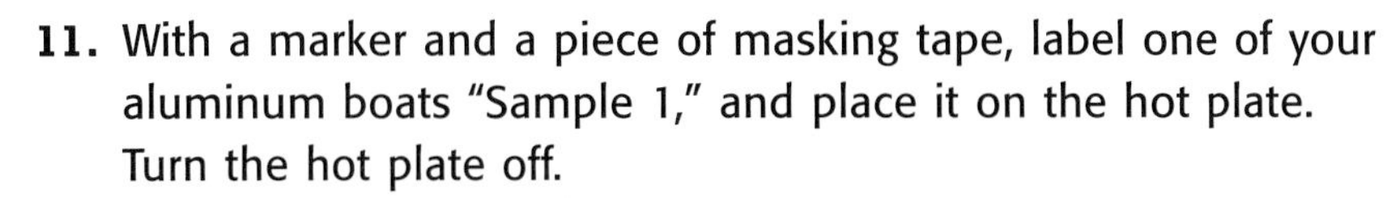

11. With a marker and a piece of masking tape, label one of your aluminum boats "Sample 1," and place it on the hot plate. Turn the hot plate off.

12. Label the other aluminum boat "Sample 2," and place it on the lab table.

13. Using the test-tube tongs, remove the test tube from the beaker of water, and evenly distribute the contents to each of your foil boats. Carefully pour the hot water in the beaker down the drain. Do not move or disturb either of your foil boats.

Make Observations

14. Copy the table below into your ScienceLog. Using the magnifying lens, carefully observe the foil boats. Record the time it takes for the first crystals to appear.

Crystal-Formation Table			
Crystal formation	**Time**	**Size and appearance of crystals**	**Sketch of crystals**
Sample 1			
Sample 2			

15. If crystals have not formed in the boats before class is over, carefully place the boats in a safe place. You may then record the time in days instead of in minutes.

16. When crystals have formed in both boats, use your magnifying lens to examine the crystals carefully.

Analyze the Results

17. Was your prediction correct? Explain.

18. Compare the size and shape of the crystals in Samples 1 and 2 with the size and shape of the crystals you examined in step 4. How long do you think the formation of the original crystals must have taken?

Draw Conclusions

19. Granite, basalt, and pumice are all igneous rocks. The most distinctive feature of each is the size of their crystals. Different igneous rocks form when magma cools at different rates. Examine a sample of each with your magnifying lens.

20. Copy the table below into your ScienceLog, and sketch each rock sample.

21. Use what you have learned in this activity to explain how each rock sample formed and how long it took for the crystals to form. Record your answers in your table.

Igneous Rock Observations			
	Granite	Basalt	Pumice
Sketch			
How did the rock sample form?			
Rate of cooling			

Going Further

Describe the size and shape of the crystals you would expect to find when a volcano erupts and sends material into the air and when magma oozes down the volcano's slope.

Metamorphic Mash

Metamorphism is a complex process that takes place deep within the Earth, where the temperature and pressure would turn a human into a crispy pancake. The effects of this extreme temperature and pressure are obvious in some metamorphic rocks. One of these effects is the reorganization of mineral grains within the rock. In this activity, you will investigate the process of metamorphism without being charred, flattened, or buried.

Materials

- modeling clay
- sequins or other small flat objects
- plastic knife
- small pieces of very stiff cardboard or plywood

Procedure

1. Flatten the clay into a layer about 1 cm thick. Sprinkle the surface with sequins.
2. Roll the corners of the clay toward the middle to form a neat ball.
3. Carefully use the plastic knife to cut the ball in half. In your ScienceLog, describe the position and location of the sequins inside the ball.
4. Put the ball back together, and use the sheets of cardboard or plywood to flatten the ball until it is about 2 cm thick.
5. Using the plastic knife, slice open the slab of clay in several places. In your ScienceLog, describe the position and location of the sequins in the slab.

Analysis

6. What physical process does flattening the ball represent?
7. Describe any changes in the position and location of the sequins that occurred as the clay ball was flattened into a slab.
8. How are the sequins oriented in relation to the force you put on the ball to flatten it?
9. Do you think the orientation of the mineral grains in a foliated metamorphic rock tells you anything about the rock? Defend your answer.

Going Further

Suppose you find a foliated metamorphic rock that has grains running in two distinct directions. Use what you have learned in this activity to offer a possible explanation for this observation.

Let's Get Sedimental

How do we determine if sedimentary rock layers are undisturbed? The best way is to be sure that the top of the layer still points up. This activity will show you how to read rock features that say, in effect, "This side up." Then you can look for the signs at a real outcrop.

Materials

- sand
- gravel
- soil (clay-rich, if available)
- 3 L mixing bowl
- plastic pickle jar or 3 L plastic soda bottle with a cap
- water
- scissors
- dropper pipet
- magnifying lens

Procedure

1. Thoroughly mix the sand, gravel, and soil together, and fill the plastic container about one-third full of the mixture.
2. Add water until the container is two-thirds full. Twist the cap back onto the container, and shake the container vigorously until all of the sediment is mixed in the rapidly moving water.
3. Place the container on a tabletop. Using the scissors, carefully cut the top off the container a few centimeters above the water, as shown at right. This will promote evaporation.
4. Do not disturb the container. Allow the water to evaporate. (You may accelerate the process by carefully using the dropper pipet to siphon off some of the clear water after allowing the container to sit for at least 24 hours.)
5. Immediately after you set the bottle on the desk, describe what you see from above and through the sides of the bottle. Do this at least once each day. Record your observations in your ScienceLog.
6. After the sediment has dried and hardened, describe its surface in your ScienceLog.
7. Carefully lay the container on its side, and cut a strip of plastic out of the side to expose the sediments in the bottle. You may find it easier if you place pieces of clay on either side of the bottle to stabilize it.

8. Brush away the loose material from the sediment, and gently blow on the surface until it is clean. Examine the surface, and record your observations in your ScienceLog.

Analysis

9. Do you see anything through the side of the bottle that could help you determine if a sedimentary rock is undisturbed? Explain.
10. What structures do you see on the surface of the sediment that you would not expect to find at the bottom?
11. Explain how these features might be used to identify the top of the sedimentary bed in a real outcrop and to decide if the bed has been disturbed.
12. Did you see any structures on the side of the container that might indicate which direction is up?
13. After removing the side of the bottle, use the magnifying lens to examine the boundaries between the gravel, sand, and silt. What do you see? Do the size and type of sediment change quickly or gradually?

Going Further

Explain why the following statement is true: "If the top of a layer can't be found, finding the bottom of it works just as well."

Imagine that a layer was deposited directly above the layers in your container. Describe the bottom of this layer.

Earthquake Waves

The energy from an earthquake travels as seismic waves in all directions through the Earth. Seismologists can use the properties of certain types of seismic waves to find the epicenter of an earthquake.

P waves travel more quickly than S waves and are always detected first. The average speed of P waves in the Earth's crust is 6.1 km/s. The average speed of S waves in the Earth's crust is 4.1 km/s. The difference in arrival time between P waves and S waves is called *lag time.*

In this activity you will use the S-P-time method to determine the location of an earthquake's epicenter.

Materials

- calculator (optional)
- compass
- metric ruler

Procedure

1. The illustration below shows seismographic records made in three cities following an earthquake. These traces begin at the left and show the arrival of P waves at time zero. The second set of waves on each record represents the arrival of S waves.

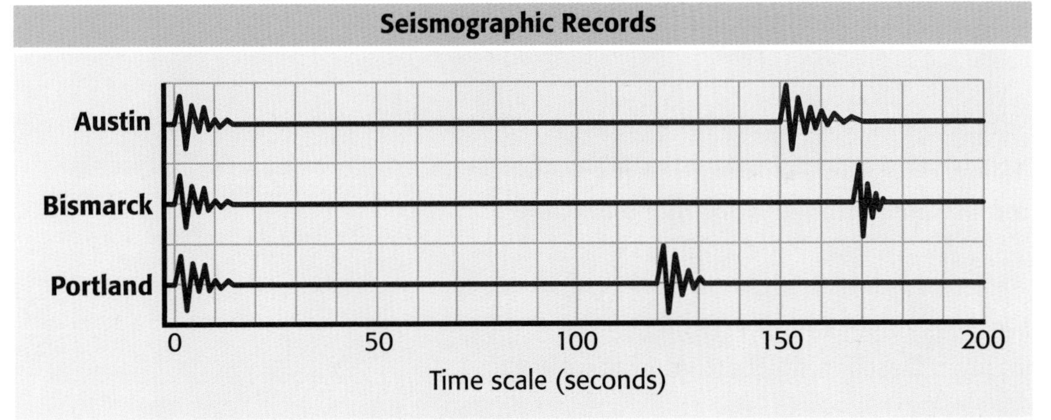

2. Copy the data table on the next page into your ScienceLog.
3. Use the time scale provided with the seismographic records to find the lag time between the P waves and the S waves for each city. Remember, the lag time is the time between the moment when the first P wave arrives and the moment when the first S wave arrives. Record this data in your table.
4. Use the following equation to calculate how long it takes each wave type to travel 100 km:

   100 km ÷ average speed of the wave = time

5. To find lag time for earthquake waves at 100 km, subtract the time it takes P waves to travel 100 km from the time it takes S waves to travel 100 km. Record the lag time in your ScienceLog.

6. Use the following formula to find the distance from each city to the epicenter:

$$\text{distance} = \frac{\text{measured lag time (s)} \times 100\text{ km}}{\text{lag time for 100 km (s)}}$$

In your Data Table, record the distance from each city to the epicenter.

Epicenter Data Table		
City	Lag time (seconds)	Distance to the epicenter (km)
Austin, TX		
Bismarck, ND		
Portland, OR		

7. Trace the map below into your ScienceLog.

8. Use the scale to adjust your compass so that the radius of a circle with Austin at the center is equal to the distance between Austin and the epicenter of the earthquake.

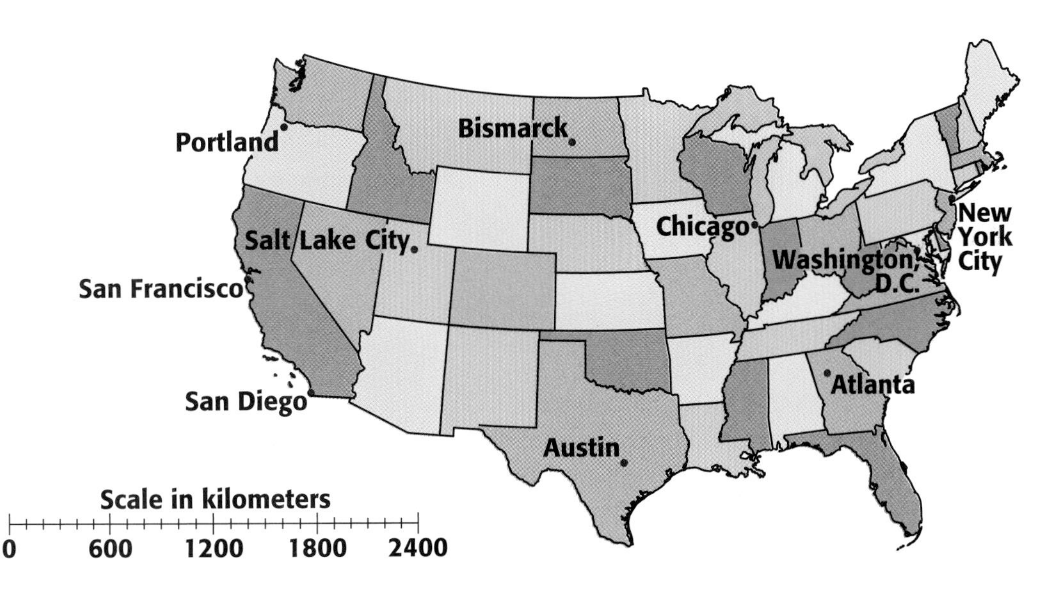

9. Put the point of your compass at Austin on your copy of the map, and draw a circle.

10. Repeat steps 8 and 9 for Bismarck and Portland. The epicenter of the earthquake is located near the point where the three circles meet.

Analysis

11. Which city is closest to the epicenter?

12. Why do seismologists need measurements from three different locations to find the epicenter of an earthquake?

Volcano Verdict

You will need to pair up with a partner for this exploration. You and your partner will act as geologists who work in a city located near a volcano. City officials are counting on you to predict when the volcano will erupt next. You and your partner have decided to use limewater as a gas-emissions tester. You will use this tester to measure the levels of carbon dioxide emitted from a simulated volcano. The more active the volcano is, the more carbon dioxide it releases.

Materials

- 1 L of limewater
- 9 oz clear plastic cup
- graduated cylinder
- 100 mL of water
- 140 mL of white vinegar
- 16 oz drink bottle
- modeling clay
- flexible drinking straw
- 15 mL of baking soda
- 2 sheets of bathroom tissue
- coin
- box or stand for plastic cup

Procedure

1. Put on your safety goggles, and carefully pour limewater into the plastic cup until the cup is three-fourths full. This is your gas-emissions tester.
2. Now build a model volcano. Begin by pouring 50 mL of water and 70 mL of vinegar into the drink bottle.
3. Form a plug of clay around the short end of the straw, as shown below. The clay plug must be large enough to cover the opening of the bottle. Be careful not to get the clay wet.
4. Sprinkle 5 mL of baking soda along the center of a single section of bathroom tissue. Then roll the tissue and twist the ends so that the baking soda can't fall out.

5. Drop the tissue into the drink bottle, and immediately put the short end of the straw inside the bottle, making a seal with the clay.
6. Put the other end of the straw into the limewater, as shown at right.
7. You have just taken your first measurement of gas levels from the volcano. Record your observations in your ScienceLog.
8. Imagine that it is several days later and you need to test the volcano again to collect more data. Before you continue, toss a coin. If it lands heads up, go to step 9a. If it lands tails up, go to step 9b. Write the step you take in your ScienceLog.

9a. Repeat steps 1–7. This time add 2 mL of baking soda to the vinegar and water. **Note:** You must use fresh water, vinegar, and limewater. Describe your observations in your ScienceLog. Go to step 10.

9b. Repeat steps 1–7. This time add 8 mL of baking soda to the vinegar and water. **Note:** You must use fresh water, vinegar, and limewater. Describe your observations in your ScienceLog. Go to step 10.

Analysis

10. How do you explain the difference in the appearance of the limewater from one trial to the next?
11. What do your measurements indicate about the activity in the volcano?
12. Based on your results, do you think it would be necessary to evacuate the city?
13. How would a geologist use a gas-emissions tester to forecast volcanic eruptions?

Concept Mapping: A Way to Bring Ideas Together

What Is a Concept Map?

Have you ever tried to tell someone about a book or a chapter you've just read and found that you can remember only a few isolated words and ideas? Or maybe you've memorized facts for a test and then weeks later discovered you're not even sure what topics those facts covered.

In both cases, you may have understood the ideas or concepts by themselves but not in relation to one another. If you could somehow link the ideas together, you would probably understand them better and remember them longer. This is something a concept map can help you do. A concept map is a way to see how ideas or concepts fit together. It can help you see the "big picture."

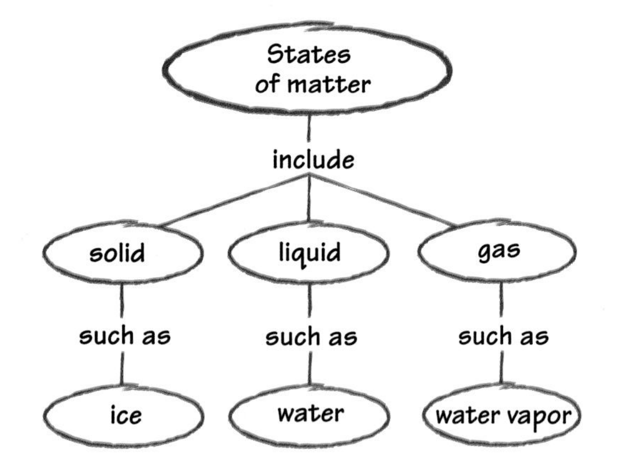

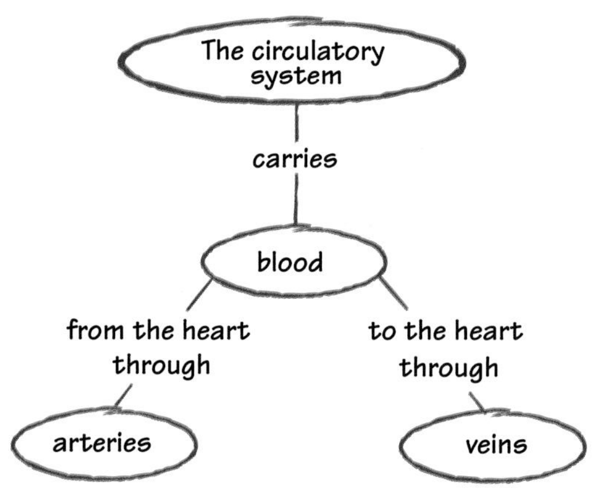

How to Make a Concept Map

1. **Make a list of the main ideas or concepts.**

   It might help to write each concept on its own slip of paper. This will make it easier to rearrange the concepts as many times as necessary to make sense of how the concepts are connected. After you've made a few concept maps this way, you can go directly from writing your list to actually making the map.

2. **Arrange the concepts in order from the most general to the most specific.**

   Put the most general concept at the top and circle it. Ask yourself, "How does this concept relate to the remaining concepts?" As you see the relationships, arrange the concepts in order from general to specific.

3. **Connect the related concepts with lines.**

4. **On each line, write an action word or short phrase that shows how the concepts are related.**

   Look at the concept maps on this page, and then see if you can make one for the following terms:

   plants, water, photosynthesis, carbon dioxide, sun's energy

   One possible answer is provided at right, but don't look at it until you try the concept map yourself.

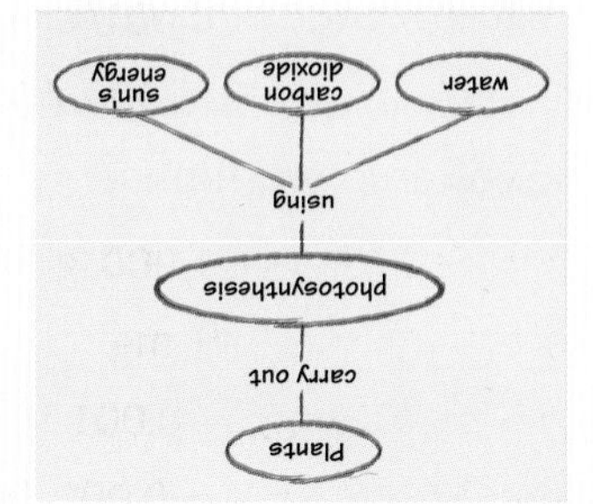

SI Measurement

The International System of Units, or SI, is the standard system of measurement used by many scientists. Using the same standards of measurement makes it easier for scientists to communicate with one another.

SI works by combining prefixes and base units. Each base unit can be used with different prefixes to define smaller and larger quantities. The table below lists common SI prefixes.

SI Prefixes

Prefix	Abbreviation	Factor	Example
kilo-	k	1,000	kilogram, 1 kg = 1,000 g
hecto-	h	100	hectoliter, 1 hL = 100 L
deka-	da	10	dekameter, 1 dam = 10 m
		1	meter, liter
deci-	d	0.1	decigram, 1 dg = 0.1 g
centi-	c	0.01	centimeter, 1 cm = 0.01 m
milli-	m	0.001	milliliter, 1 mL = 0.001 L
micro-	µ	0.000 001	micrometer, 1 µm = 0.000 001 m

SI Conversion Table

SI units	From SI to English	From English to SI
Length		
kilometer (km) = 1,000 m	1 km = 0.621 mi	1 mi = 1.609 km
meter (m) = 100 cm	1 m = 3.281 ft	1 ft = 0.305 m
centimeter (cm) = 0.01 m	1 cm = 0.394 in.	1 in. = 2.540 cm
millimeter (mm) = 0.001 m	1 mm = 0.039 in.	
micrometer (µm) = 0.000 001 m		
nanometer (nm) = 0.000 000 001 m		
Area		
square kilometer (km^2) = 100 hectares	1 km^2 = 0.386 mi^2	1 mi^2 = 2.590 km^2
hectare (ha) = 10,000 m^2	1 ha = 2.471 acres	1 acre = 0.405 ha
square meter (m^2) = 10,000 cm^2	1 m^2 = 10.765 ft^2	1 ft^2 = 0.093 m^2
square centimeter (cm^2) = 100 mm^2	1 cm^2 = 0.155 $in.^2$	1 $in.^2$ = 6.452 cm^2
Volume		
liter (L) = 1,000 mL = 1 dm^3	1 L = 1.057 fl qt	1 fl qt = 0.946 L
milliliter (mL) = 0.001 L = 1 cm^3	1 mL = 0.034 fl oz	1 fl oz = 29.575 mL
microliter (µL) = 0.000 001 L		
Mass		
kilogram (kg) = 1,000 g	1 kg = 2.205 lb	1 lb = 0.454 kg
gram (g) = 1,000 mg	1 g = 0.035 oz	1 oz = 28.349 g
milligram (mg) = 0.001 g		
microgram (µg) = 0.000 001 g		

Temperature Scales

Temperature can be expressed using three different scales: Fahrenheit, Celsius, and Kelvin. The SI unit for temperature is the kelvin (K).

Although 0 K is much colder than 0°C, a change of 1 K is equal to a change of 1°C.

Three Temperature Scales

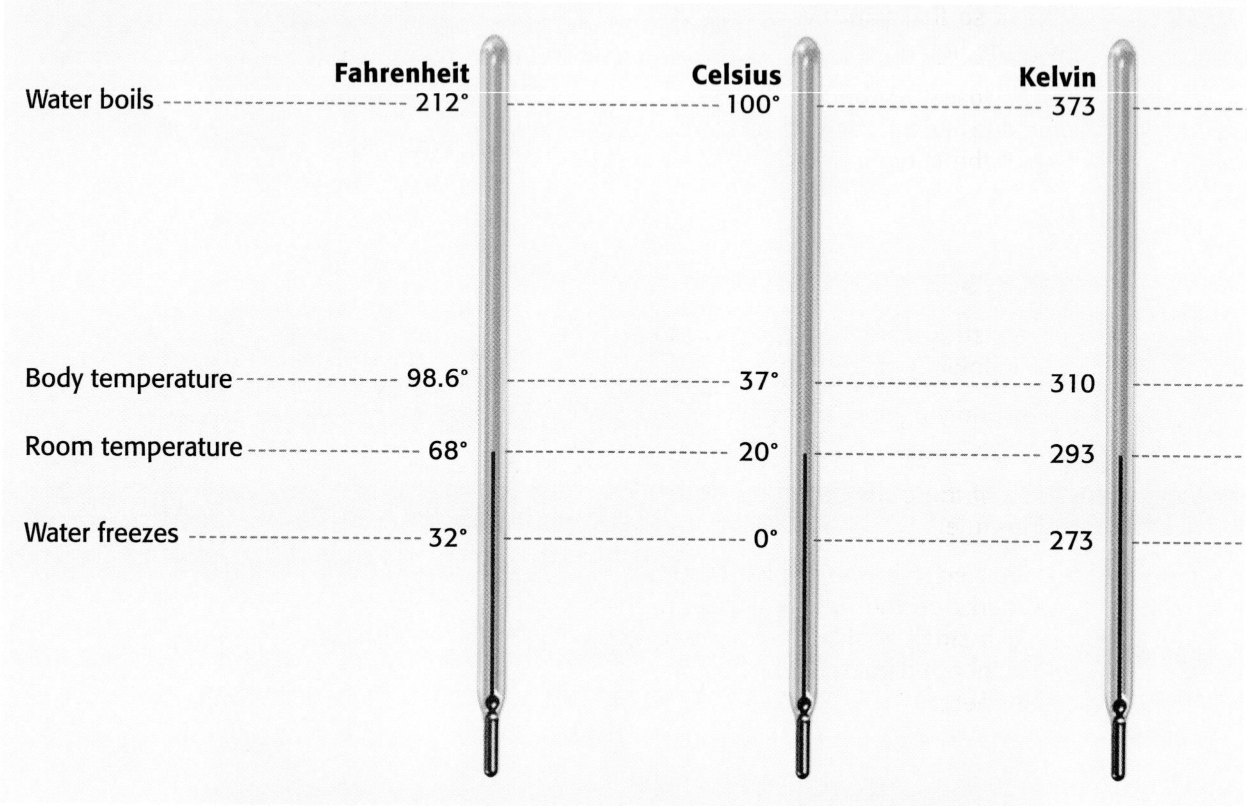

	Fahrenheit	Celsius	Kelvin
Water boils	212°	100°	373
Body temperature	98.6°	37°	310
Room temperature	68°	20°	293
Water freezes	32°	0°	273

Temperature Conversions Table

To convert	Use this equation:	Example
Celsius to Fahrenheit °C → °F	$°F = \left(\frac{9}{5} \times °C\right) + 32$	Convert 45°C to °F. $°F = \left(\frac{9}{5} \times 45°C\right) + 32 = 113°F$
Fahrenheit to Celsius °F → °C	$°C = \frac{5}{9} \times (°F - 32)$	Convert 68°F to °C. $°C = \frac{5}{9} \times (68°F - 32) = 20°C$
Celsius to Kelvin °C → K	$K = °C + 273$	Convert 45°C to K. $K = 45°C + 273 = 318\ K$
Kelvin to Celsius K → °C	$°C = K - 273$	Convert 32 K to °C. $°C = 32\ K - 273 = -241°C$

Measuring Skills

Using a Graduated Cylinder

When using a graduated cylinder to measure volume, keep the following procedures in mind:

1. Make sure the cylinder is on a flat, level surface.
2. Move your head so that your eye is level with the surface of the liquid.
3. Read the mark closest to the liquid level. On glass graduated cylinders, read the mark closest to the center of the curve in the liquid's surface.

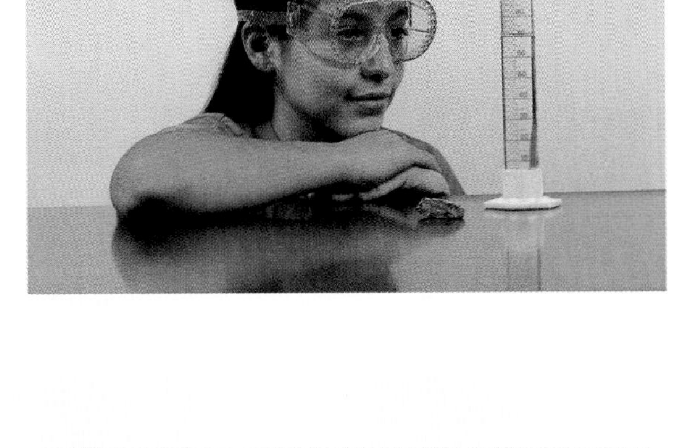

Using a Meterstick or Metric Ruler

When using a meterstick or metric ruler to measure length, keep the following procedures in mind:

1. Place the ruler firmly against the object you are measuring.
2. Align one edge of the object exactly with the zero end of the ruler.
3. Look at the other edge of the object to see which of the marks on the ruler is closest to that edge. **Note:** Each small slash between the centimeters represents a millimeter, which is one-tenth of a centimeter.

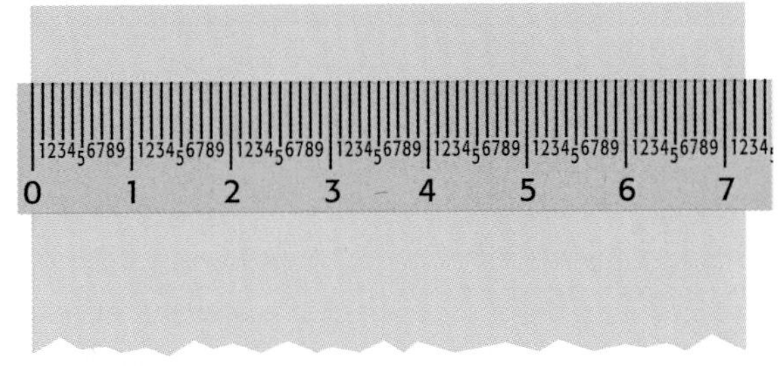

Using a Triple-Beam Balance

When using a triple-beam balance to measure mass, keep the following procedures in mind:

1. Make sure the balance is on a level surface.
2. Place all of the countermasses at zero. Adjust the balancing knob until the pointer rests at zero.
3. Place the object you wish to measure on the pan. **Caution:** Do not place hot objects or chemicals directly on the balance pan.
4. Move the largest countermass along the beam to the right until it is at the last notch that does not tip the balance. Follow the same procedure with the next-largest countermass. Then move the smallest countermass until the pointer rests at zero.
5. Add the readings from the three beams together to determine the mass of the object.
6. When determining the mass of crystals or powders, use a piece of filter paper. First find the mass of the paper. Then add the crystals or powder to the paper and remeasure. The actual mass of the crystals or powder is the total mass minus the mass of the paper. When finding the mass of liquids, first find the mass of the empty container. Then find the mass of the liquid and container together. The mass of the liquid is the total mass minus the mass of the container.

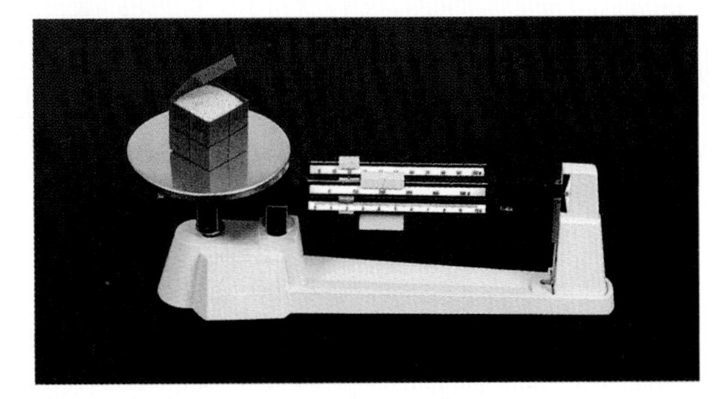

Scientific Method

The series of steps that scientists use to answer questions and solve problems is often called the **scientific method.** The scientific method is not a rigid procedure. Scientists may use all of the steps or just some of the steps of the scientific method. They may even repeat some of the steps. The goal of the scientific method is to come up with reliable answers and solutions.

Six Steps of the Scientific Method

1 **Ask a Question** Good questions come from careful **observations.** You make observations by using your senses to gather information. Sometimes you may use instruments, such as microscopes and telescopes, to extend the range of your senses. As you observe the natural world, you will discover that you have many more questions than answers. These questions drive the scientific method.

Questions beginning with *what, why, how,* and *when* are very important in focusing an investigation, and they often lead to a hypothesis. (You will learn what a hypothesis is in the next step.) Here is an example of a question that could lead to further investigation.

Question: How does acid rain affect plant growth?

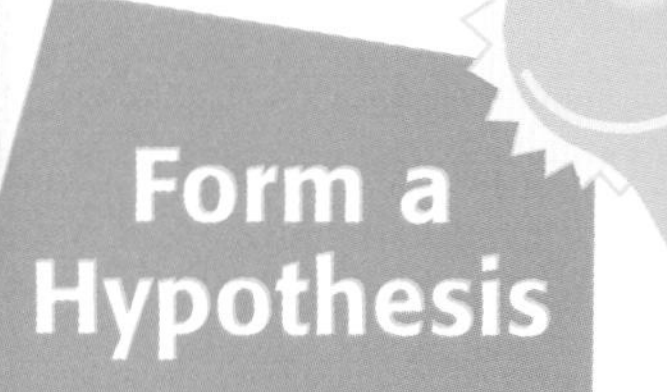

2 **Form a Hypothesis** After you come up with a question, you need to turn the question into a **hypothesis.** A hypothesis is a clear statement of what you expect the answer to your question to be. Your hypothesis will represent your best "educated guess" based on your observations and what you already know. A good hypothesis is testable. If observations and information cannot be gathered or if an experiment cannot be designed to test your hypothesis, it is untestable, and the investigation can go no further.

Here is a hypothesis that could be formed from the question, "How does acid rain affect plant growth?"

Hypothesis: Acid rain causes plants to grow more slowly.

Notice that the hypothesis provides some specifics that lead to methods of testing. The hypothesis can also lead to predictions. A **prediction** is what you think will be the outcome of your experiment or data collection. Predictions are usually stated in an "if . . . then" format. For example, **if** meat is kept at room temperature, **then** it will spoil faster than meat kept in the refrigerator. More than one prediction can be made for a single hypothesis. Here is a sample prediction for the hypothesis that acid rain causes plants to grow more slowly.

Prediction: If a plant is watered with only acid rain (which has a pH of 4), then the plant will grow at half its normal rate.

3 **Test the Hypothesis** After you have formed a hypothesis and made a prediction, you should test your hypothesis. There are different ways to do this. Perhaps the most familiar way is to conduct a **controlled experiment.** A controlled experiment tests only one factor at a time. A controlled experiment has a **control group** and one or more **experimental groups.** All the factors for the control and experimental groups are the same except for one factor, which is called the **variable.** By changing only one factor, you can see the results of just that one change.

Test the Hypothesis

Sometimes, the nature of an investigation makes a controlled experiment impossible. For example, dinosaurs have been extinct for millions of years, and the Earth's core is surrounded by thousands of meters of rock. It would be difficult, if not impossible, to conduct controlled experiments on such things. Under such circumstances, a hypothesis may be tested by making detailed observations. Taking measurements is one way of making observations.

4 **Analyze the Results** After you have completed your experiments, made your observations, and collected your data, you must analyze all the information you have gathered. Tables and graphs are often used in this step to organize the data.

5 **Draw Conclusions** Based on the analysis of your data, you should conclude whether or not your results support your hypothesis. If your hypothesis is supported, you (or others) might want to repeat the observations or experiments to verify your results. If your hypothesis is not supported by the data, you may have to check your procedure for errors. You may even have to reject your hypothesis and make a new one. If you cannot draw a conclusion from your results, you may have to try the investigation again or carry out further observations or experiments.

Draw Conclusions

Do they support your hypothesis?

No

Yes

6 **Communicate Results** After any scientific investigation, you should report your results. By doing a written or oral report, you let others know what you have learned. They may want to repeat your investigation to see if they get the same results. Your report may even lead to another question, which in turn may lead to another investigation.

Scientific Method in Action

The scientific method is not a "straight line" of steps. It contains loops in which several steps may be repeated over and over again, while others may not be necessary. For example, sometimes scientists will find that testing one hypothesis raises new questions and new hypotheses to be tested. And sometimes, testing the hypothesis leads directly to a conclusion. Furthermore, the steps in the scientific method are not always used in the same order. Follow the steps in the diagram below, and see how many different directions the scientific method can take you.

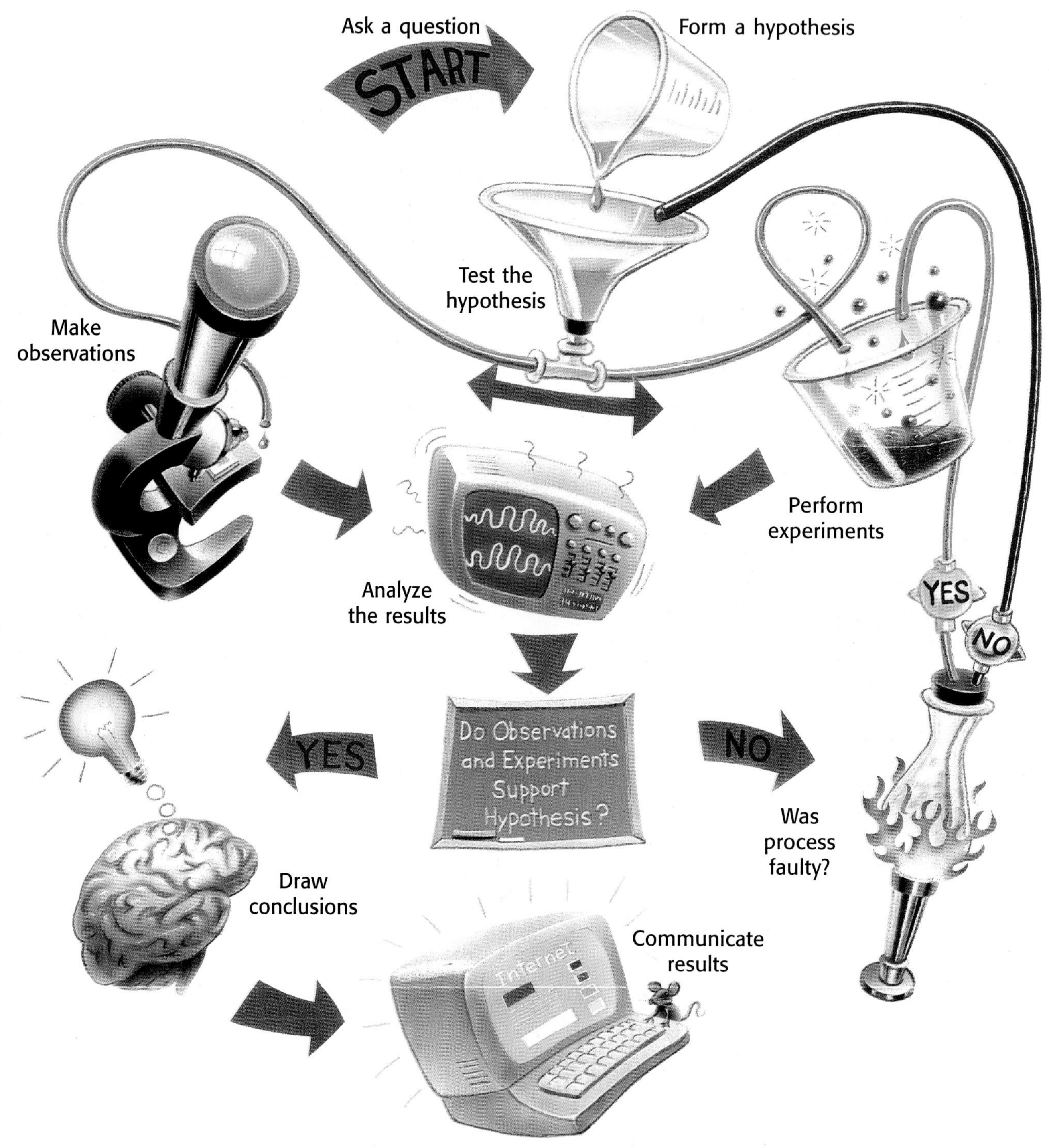

Making Charts and Graphs

Circle Graphs

A circle graph, or pie chart, shows how each group of data relates to all of the data. Each part of the circle represents a category of the data. The entire circle represents all of the data. For example, a biologist studying a hardwood forest in Wisconsin found that there were five different types of trees. The data table at right summarizes the biologist's findings.

Wisconsin Hardwood Trees

Type of tree	Number found
Oak	600
Maple	750
Beech	300
Birch	1,200
Hickory	150
Total	3,000

How to Make a Circle Graph

1. In order to make a circle graph of this data, first find the percentage of each type of tree. To do this, divide the number of individual trees by the total number of trees and multiply by 100.

$\frac{\text{600 oak}}{\text{3,000 trees}} \times 100 = 20\%$

$\frac{\text{750 maple}}{\text{3,000 trees}} \times 100 = 25\%$

$\frac{\text{300 beech}}{\text{3,000 trees}} \times 100 = 10\%$

$\frac{\text{1,200 birch}}{\text{3,000 trees}} \times 100 = 40\%$

$\frac{\text{150 hickory}}{\text{3,000 trees}} \times 100 = 5\%$

2. Now determine the size of the pie shapes that make up the chart. Do this by multiplying each percentage by 360°. Remember that a circle contains 360°.

$20\% \times 360° = 72°$ $\quad$ $25\% \times 360° = 90°$

$10\% \times 360° = 36°$ $\quad$ $40\% \times 360° = 144°$

$5\% \times 360° = 18°$

3. Then check that the sum of the percentages is 100 and the sum of the degrees is 360.

$20\% + 25\% + 10\% + 40\% + 5\% = 100\%$

$72° + 90° + 36° + 144° + 18° = 360°$

4. Use a compass to draw a circle and mark its center.

5. Then use a protractor to draw angles of 72°, 90°, 36°, 144°, and 18° in the circle.

6. Finally, label each part of the graph, and choose an appropriate title.

A Community of Wisconsin Hardwood Trees

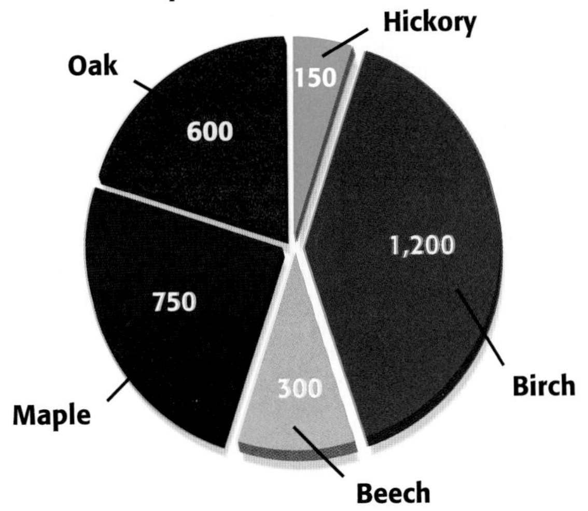

APPENDIX

Line Graphs

Population of Appleton, 1900–2000	
Year	**Population**
1900	1,800
1920	2,500
1940	3,200
1960	3,900
1980	4,600
2000	5,300

Line graphs are most often used to demonstrate continuous change. For example, Mr. Smith's science class analyzed the population records for their hometown, Appleton, between 1900 and 2000. Examine the data at left.

Because the year and the population change, they are the *variables*. The population is determined by, or dependent on, the year. Therefore, the population is called the **dependent variable**, and the year is called the **independent variable.** Each set of data is called a **data pair.** To prepare a line graph, data pairs must first be organized in a table like the one at left.

How to Make a Line Graph

1. Place the independent variable along the horizontal (x) axis. Place the dependent variable along the vertical (y) axis.
2. Label the x-axis "Year" and the y-axis "Population." Look at your largest and smallest values for the population. Determine a scale for the y-axis that will provide enough space to show these values. You must use the same scale for the entire length of the axis. Find an appropriate scale for the x-axis too.
3. Choose reasonable starting points for each axis.
4. Plot the data pairs as accurately as possible.
5. Choose a title that accurately represents the data.

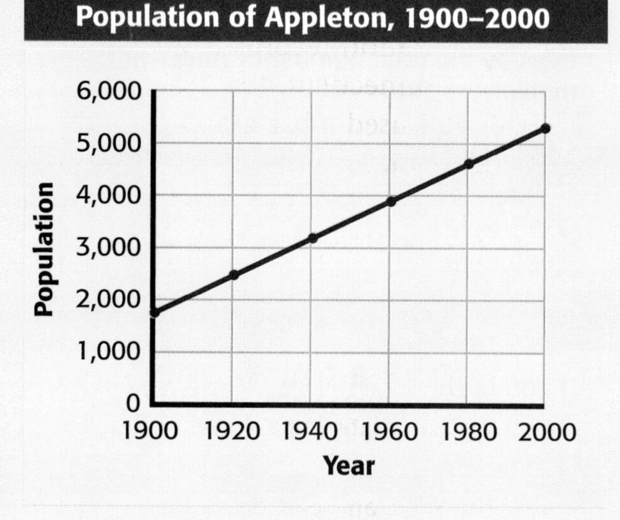

How to Determine Slope

Slope is the ratio of the change in the y-axis to the change in the x-axis, or "rise over run."

1. Choose two points on the line graph. For example, the population of Appleton in 2000 was 5,300 people. Therefore, you can define point a as (2000, 5,300). In 1900, the population was 1,800 people. Define point b as (1900, 1,800).
2. Find the change in the y-axis.
   (y at point a) – (y at point b)
   5,300 people – 1,800 people = 3,500 people
3. Find the change in the x-axis.
   (x at point a) – (x at point b)
   2000 – 1900 = 100 years
4. Calculate the slope of the graph by dividing the change in y by the change in x.

$$\text{slope} = \frac{\text{change in } y}{\text{change in } x}$$

$$\text{slope} = \frac{3{,}500 \text{ people}}{100 \text{ years}}$$

$$\text{slope} = 35 \text{ people per year}$$

In this example, the population in Appleton increased by a fixed amount each year. The graph of this data is a straight line. Therefore, the relationship is **linear.** When the graph of a set of data is not a straight line, the relationship is **nonlinear.**

Using Algebra to Determine Slope

The equation in step 4 may also be arranged to be:

$$y = kx$$

where y represents the change in the y-axis, k represents the slope, and x represents the change in the x-axis.

$$\text{slope} = \frac{\text{change in } y}{\text{change in } x}$$

$$k = \frac{y}{x}$$

$$k \times x = \frac{y \times x}{x}$$

$$kx = y$$

Bar Graphs

Bar graphs are used to demonstrate change that is not continuous. These graphs can be used to indicate trends when the data are taken over a long period of time. A meteorologist gathered the precipitation records at right for Hartford, Connecticut, for April 1–15, 1996, and used a bar graph to represent the data.

Precipitation in Hartford, Connecticut April 1–15, 1996

Date	Precipitation (cm)	Date	Precipitation (cm)
April 1	0.5	April 9	0.25
April 2	1.25	April 10	0.0
April 3	0.0	April 11	1.0
April 4	0.0	April 12	0.0
April 5	0.0	April 13	0.25
April 6	0.0	April 14	0.0
April 7	0.0	April 15	6.50
April 8	1.75		

How to Make a Bar Graph

1. Use an appropriate scale and a reasonable starting point for each axis.
2. Label the axes, and plot the data.
3. Choose a title that accurately represents the data.

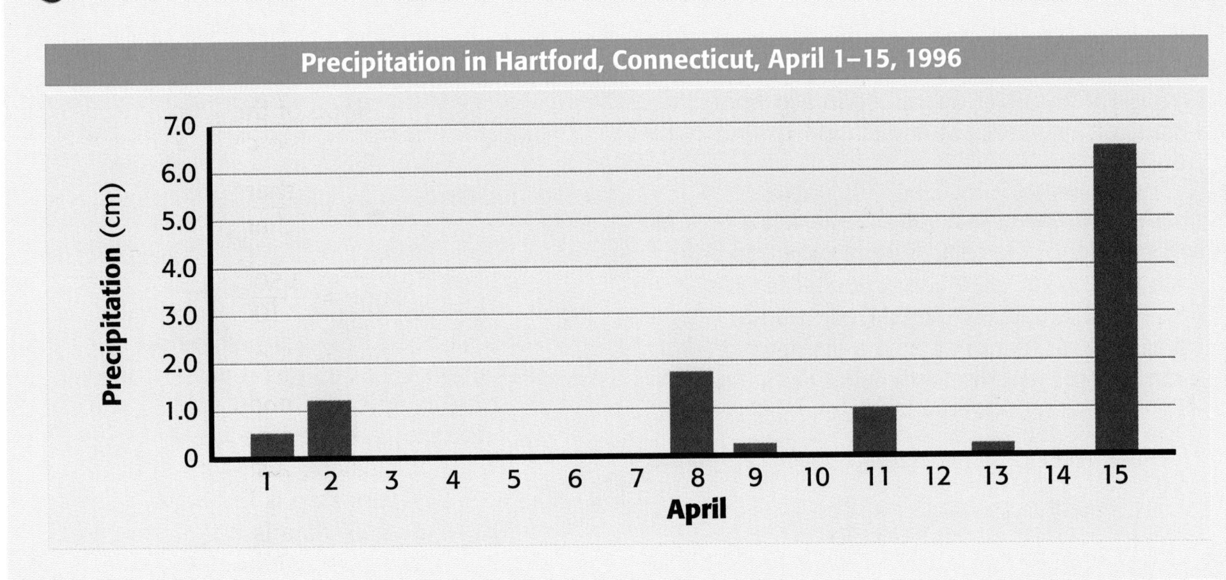

APPENDIX

Math Refresher

Science requires an understanding of many math concepts. The following pages will help you review some important math skills.

Averages

An **average,** or **mean,** simplifies a list of numbers into a single number that *approximates* their value.

Example: Find the average of the following set of numbers: 5, 4, 7, and 8.

Step 1: Find the sum.

$$5 + 4 + 7 + 8 = 24$$

Step 2: Divide the sum by the amount of numbers in your set. Because there are four numbers in this example, divide the sum by 4.

$$\frac{24}{4} = 6$$

The average, or mean, is **6.**

Ratios

A **ratio** is a comparison between numbers, and it is usually written as a fraction.

Example: Find the ratio of thermometers to students if you have 36 thermometers and 48 students in your class.

Step 1: Make the ratio.

$$\frac{36 \text{ thermometers}}{48 \text{ students}}$$

Step 2: Reduce the fraction to its simplest form.

$$\frac{36}{48} = \frac{36 \div 12}{48 \div 12} = \frac{3}{4}$$

The ratio of thermometers to students is **3 to 4,** or $\frac{3}{4}$. The ratio may also be written in the form 3:4.

Proportions

A **proportion** is an equation that states that two ratios are equal.

$$\frac{3}{1} = \frac{12}{4}$$

To solve a proportion, first multiply across the equal sign. This is called cross-multiplication. If you know three of the quantities in a proportion, you can use cross-multiplication to find the fourth.

Example: Imagine that you are making a scale model of the solar system for your science project. The diameter of Jupiter is 11.2 times the diameter of the Earth. If you are using a plastic-foam ball with a diameter of 2 cm to represent the Earth, what diameter does the ball representing Jupiter need to be?

$$\frac{11.2}{1} = \frac{x}{2 \text{ cm}}$$

Step 1: Cross-multiply.

$$\frac{11.2}{1} \times \frac{x}{2}$$

$$11.2 \times 2 = x \times 1$$

Step 2: Multiply.

$$22.4 = x \times 1$$

Step 3: Isolate the variable by dividing both sides by 1.

$$x = \frac{22.4}{1}$$

$$x = 22.4 \text{ cm}$$

You will need to use a ball with a diameter of **22.4 cm** to represent Jupiter.

Percentages

A **percentage** is a ratio of a given number to 100.

Example: What is 85 percent of 40?

Step 1: Rewrite the percentage by moving the decimal point two places to the left.

$$.85$$

Step 2: Multiply the decimal by the number you are calculating the percentage of.

$$0.85 \times 40 = 34$$

85 percent of 40 is **34.**

Decimals

To **add** or **subtract decimals,** line up the digits vertically so that the decimal points line up. Then add or subtract the columns from right to left, carrying or borrowing numbers as necessary.

Example: Add the following numbers: 3.1415 and 2.96.

Step 1: Line up the digits vertically so that the decimal points line up.

$$\begin{array}{r} 3.1415 \\ +\ 2.96 \\ \hline \end{array}$$

Step 2: Add the columns from right to left, carrying when necessary.

$$\begin{array}{r} {\scriptstyle 1\ 1} \\ 3.1415 \\ +\ 2.96 \\ \hline 6.1015 \end{array}$$

The sum is **6.1015.**

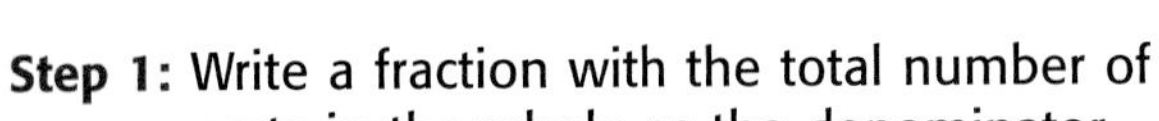

Fractions

Numbers tell you how many; **fractions** tell you *how much of a whole.*

Example: Your class has 24 plants. Your teacher instructs you to put 5 in a shady spot. What fraction does this represent?

Step 1: Write a fraction with the total number of parts in the whole as the denominator.

$$\frac{?}{24}$$

Step 2: Write the number of parts of the whole being represented as the numerator.

$$\frac{5}{24}$$

$\mathbf{\frac{5}{24}}$ of the plants will be in the shade.

Reducing Fractions

It is usually best to express a fraction in simplest form. This is called *reducing* a fraction.

Example: Reduce the fraction $\frac{30}{45}$ to its simplest form.

Step 1: Find the largest whole number that will divide evenly into both the numerator and denominator. This number is called the greatest common factor (GCF).

factors of the numerator 30: 1, 2, 3, 5, 6, 10, **15,** 30

factors of the denominator 45: 1, 3, 5, 9, **15,** 45

Step 2: Divide both the numerator and the denominator by the GCF, which in this case is 15.

$$\frac{30}{45} = \frac{30 \div 15}{45 \div 15} = \frac{2}{3}$$

$\frac{30}{45}$ reduced to its simplest form is $\mathbf{\frac{2}{3}}$.

APPENDIX

Adding and Subtracting Fractions

To **add** or **subtract fractions** that have the **same denominator,** simply add or subtract the numerators.

Examples:

$$\frac{3}{5} + \frac{1}{5} = ? \text{ and } \frac{3}{4} - \frac{1}{4} = ?$$

Step 1: Add or subtract the numerators.

$$\frac{3}{5} + \frac{1}{5} = \frac{4}{\ } \text{ and } \frac{3}{4} - \frac{1}{4} = \frac{2}{\ }$$

Step 2: Write the sum or difference over the denominator.

$$\frac{3}{5} + \frac{1}{5} = \frac{4}{5} \text{ and } \frac{3}{4} - \frac{1}{4} = \frac{2}{4}$$

Step 3: If necessary, reduce the fraction to its simplest form.

$\mathbf{\frac{4}{5}}$ cannot be reduced, and $\frac{2}{4} = \mathbf{\frac{1}{2}}$.

To **add** or **subtract fractions** that have **different denominators,** first find the least common denominator (LCD).

Examples:

$$\frac{1}{2} + \frac{1}{6} = ? \text{ and } \frac{3}{4} - \frac{2}{3} = ?$$

Step 1: Write the equivalent fractions with a common denominator.

$$\frac{3}{6} + \frac{1}{6} = ? \text{ and } \frac{9}{12} - \frac{8}{12} = ?$$

Step 2: Add or subtract.

$$\frac{3}{6} + \frac{1}{6} = \frac{4}{6} \text{ and } \frac{9}{12} - \frac{8}{12} = \frac{1}{12}$$

Step 3: If necessary, reduce the fraction to its simplest form.

$\frac{4}{6} = \mathbf{\frac{2}{3}}$, and $\mathbf{\frac{1}{12}}$ cannot be reduced.

Multiplying Fractions

To **multiply fractions,** multiply the numerators and the denominators together, and then reduce the fraction to its simplest form.

Example:

$$\frac{5}{9} \times \frac{7}{10} = ?$$

Step 1: Multiply the numerators and denominators.

$$\frac{5}{9} \times \frac{7}{10} = \frac{5 \times 7}{9 \times 10} = \frac{35}{90}$$

Step 2: Reduce.

$$\frac{35}{90} = \frac{35 \div 5}{90 \div 5} = \mathbf{\frac{7}{18}}$$

Dividing Fractions

To **divide fractions**, first rewrite the divisor (the number you divide *by*) upside down. This is called the reciprocal of the divisor. Then you can multiply and reduce if necessary.

Example:

$$\frac{5}{8} \div \frac{3}{2} = ?$$

Step 1: Rewrite the divisor as its reciprocal.

$$\frac{3}{2} \rightarrow \frac{2}{3}$$

Step 2: Multiply.

$$\frac{5}{8} \times \frac{2}{3} = \frac{5 \times 2}{8 \times 3} = \frac{10}{24}$$

Step 3: Reduce.

$$\frac{10}{24} = \frac{10 \div 2}{24 \div 2} = \mathbf{\frac{5}{12}}$$

Scientific Notation

Scientific notation is a short way of representing very large and very small numbers without writing all of the place-holding zeros.

Example: Write 653,000,000 in scientific notation.

Step 1: Write the number without the place-holding zeros.

653

Step 2: Place the decimal point after the first digit.

6.53

Step 3: Find the exponent by counting the number of places that you moved the decimal point.

The decimal point was moved eight places to the left. Therefore, the exponent of 10 is positive 8. Remember, if the decimal point had moved to the right, the exponent would be negative.

Step 4: Write the number in scientific notation.

$\mathbf{6.53 \times 10^8}$

Area

Area is the number of square units needed to cover the surface of an object.

Formulas:

Area of a square = side × side

Area of a rectangle = length × width

Area of a triangle = $\frac{1}{2}$ × base × height

Examples: Find the areas.

4 cm

3 cm

Triangle

Area = $\frac{1}{2}$ × base × height

Area = $\frac{1}{2}$ × 3 cm × 4 cm

Area = **6 cm²**

3 cm

6 cm

Rectangle

Area = length × width

Area = 6 cm × 3 cm

Area = **18 $\mathbf{cm^2}$**

3 cm

3 cm

Square

Area = side × side

Area = 3 cm × 3 cm

Area = **9 $\mathbf{cm^2}$**

Volume

Volume is the amount of space something occupies.

Formulas:

Volume of a cube = side × side × side

Volume of a prism = area of base × height

Examples:

Find the volume of the solids.

Cube

Volume = side × side × side

Volume = 4 cm × 4 cm × 4 cm

Volume = **64 $\mathbf{cm^3}$**

4 cm

4 cm

4 cm

4 cm

3 cm

5 cm

Prism

Volume = area of base × height

Volume = (area of triangle) × height

Volume = $\left(\frac{1}{2} \times 3 \text{ cm} \times 4 \text{ cm}\right) \times 5 \text{ cm}$

Volume = 6 cm^2 × 5 cm

Volume = **30 $\mathbf{cm^3}$**

Periodic Table of the Elements

Each square on the table includes an element's name, chemical symbol, atomic number, and atomic mass.

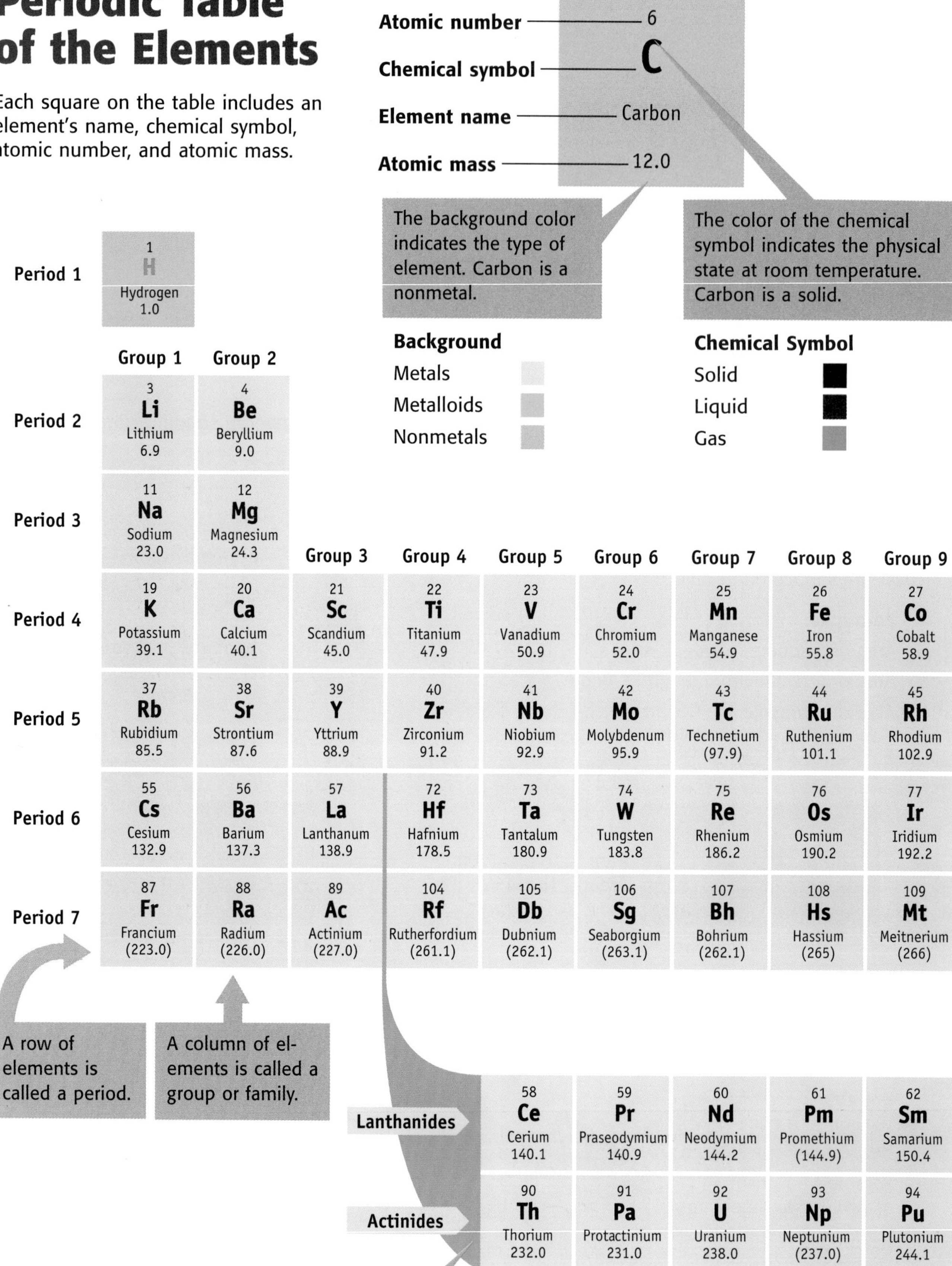

	Group 1	Group 2	Group 3	Group 4	Group 5	Group 6	Group 7	Group 8	Group 9
Period 1	1 H Hydrogen 1.0								
Period 2	3 Li Lithium 6.9	4 Be Beryllium 9.0							
Period 3	11 Na Sodium 23.0	12 Mg Magnesium 24.3							
Period 4	19 K Potassium 39.1	20 Ca Calcium 40.1	21 Sc Scandium 45.0	22 Ti Titanium 47.9	23 V Vanadium 50.9	24 Cr Chromium 52.0	25 Mn Manganese 54.9	26 Fe Iron 55.8	27 Co Cobalt 58.9
Period 5	37 Rb Rubidium 85.5	38 Sr Strontium 87.6	39 Y Yttrium 88.9	40 Zr Zirconium 91.2	41 Nb Niobium 92.9	42 Mo Molybdenum 95.9	43 Tc Technetium (97.9)	44 Ru Ruthenium 101.1	45 Rh Rhodium 102.9
Period 6	55 Cs Cesium 132.9	56 Ba Barium 137.3	57 La Lanthanum 138.9	72 Hf Hafnium 178.5	73 Ta Tantalum 180.9	74 W Tungsten 183.8	75 Re Rhenium 186.2	76 Os Osmium 190.2	77 Ir Iridium 192.2
Period 7	87 Fr Francium (223.0)	88 Ra Radium (226.0)	89 Ac Actinium (227.0)	104 Rf Rutherfordium (261.1)	105 Db Dubnium (262.1)	106 Sg Seaborgium (263.1)	107 Bh Bohrium (262.1)	108 Hs Hassium (265)	109 Mt Meitnerium (266)

Lanthanides	58 Ce Cerium 140.1	59 Pr Praseodymium 140.9	60 Nd Neodymium 144.2	61 Pm Promethium (144.9)	62 Sm Samarium 150.4
Actinides	90 Th Thorium 232.0	91 Pa Protactinium 231.0	92 U Uranium 238.0	93 Np Neptunium (237.0)	94 Pu Plutonium 244.1

This zigzag line reminds you where the metals, nonmetals, and metalloids are.

Group 10	Group 11	Group 12	Group 13	Group 14	Group 15	Group 16	Group 17	Group 18
								2 **He** Helium 4.0
			5 **B** Boron 10.8	6 **C** Carbon 12.0	7 **N** Nitrogen 14.0	8 **O** Oxygen 16.0	9 **F** Fluorine 19.0	10 **Ne** Neon 20.2
			13 **Al** Aluminum 27.0	14 **Si** Silicon 28.1	15 **P** Phosphorus 31.0	16 **S** Sulfur 32.1	17 **Cl** Chlorine 35.5	18 **Ar** Argon 39.9
28 **Ni** Nickel 58.7	29 **Cu** Copper 63.5	30 **Zn** Zinc 65.4	31 **Ga** Gallium 69.7	32 **Ge** Germanium 72.6	33 **As** Arsenic 74.9	34 **Se** Selenium 79.0	35 **Br** Bromine 79.9	36 **Kr** Krypton 83.8
46 **Pd** Palladium 106.4	47 **Ag** Silver 107.9	48 **Cd** Cadmium 112.4	49 **In** Indium 114.8	50 **Sn** Tin 118.7	51 **Sb** Antimony 121.8	52 **Te** Tellurium 127.6	53 **I** Iodine 126.9	54 **Xe** Xenon 131.3
78 **Pt** Platinum 195.1	79 **Au** Gold 197.0	80 **Hg** Mercury 200.6	81 **Tl** Thallium 204.4	82 **Pb** Lead 207.2	83 **Bi** Bismuth 209.0	84 **Po** Polonium (209.0)	85 **At** Astatine (210.0)	86 **Rn** Radon (222.0)
110 **Uun*** Ununnilium (271)	111 **Uuu*** Unununium (272)	112 **Uub*** Ununbium (277)		114 **Uuq*** Ununquadium (285)		116 **Uuh*** Ununhexium (289)		118 **Uuo*** Ununoctium (293)

A number in parenthesis is the mass number of the most stable form of that element.

63 **Eu** Europium 152.0	64 **Gd** Gadolinium 157.3	65 **Tb** Terbium 158.9	66 **Dy** Dysprosium 162.5	67 **Ho** Holmium 164.9	68 **Er** Erbium 167.3	69 **Tm** Thulium 168.9	70 **Yb** Ytterbium 173.0	71 **Lu** Lutetium 175.0
95 **Am** Americium (243.1)	96 **Cm** Curium (247.1)	97 **Bk** Berkelium (247.1)	98 **Cf** Californium (251.1)	99 **Es** Einsteinium (252.1)	100 **Fm** Fermium (257.1)	101 **Md** Mendelevium (258.1)	102 **No** Nobelium (259.1)	103 **Lr** Lawrencium (262.1)

**The official names and symbols for the elements greater than 109 will eventually be approved by a committee of scientists.*

Physical Science Refresher

Atoms and Elements

Every object in the universe is made up of particles of some kind of matter. **Matter** is anything that takes up space and has mass. All matter is made up of elements. An **element** is a substance that cannot be separated into simpler components by ordinary chemical means. This is because each element consists of only one kind of atom. An **atom** is the smallest unit of an element that has all of the properties of that element.

Atomic Structure

Atoms are made up of small particles called subatomic particles. The three major types of subatomic particles are **electrons, protons,** and **neutrons.** Electrons have a negative electric charge, protons have a positive charge, and neutrons have no electric charge. The protons and neutrons are packed close to one another to form the **nucleus.** The protons give the nucleus a positive charge. Electrons are most likely to be found in regions around the nucleus called **electron clouds.** The negatively charged electrons are attracted to the positively charged nucleus. An atom may have several energy levels in which electrons are located.

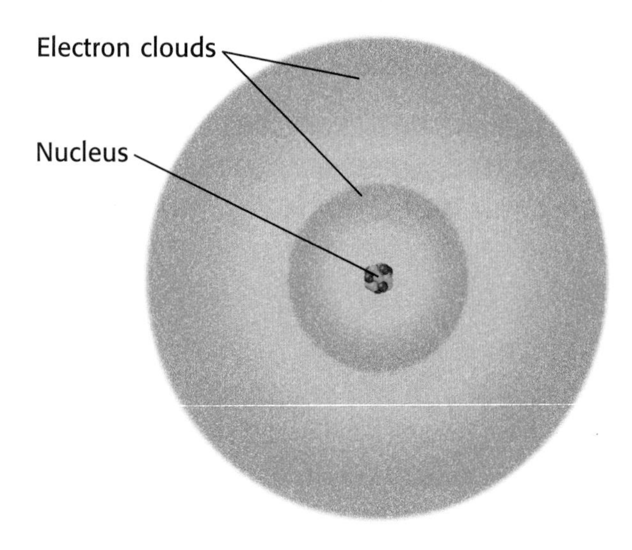

Atomic Number

To help in the identification of elements, scientists have assigned an **atomic number** to each kind of atom. The atomic number is the number of protons in the atom. Atoms with the same number of protons are all the same kind of element. In an uncharged, or electrically neutral, atom there are an equal number of protons and electrons. Therefore, the atomic number equals the number of electrons in an uncharged atom. The number of neutrons, however, can vary for a given element. Atoms of the same element that have different numbers of neutrons are called **isotopes.**

Periodic Table of the Elements

In the periodic table, the elements are arranged from left to right in order of increasing atomic number. Each element in the table is in a separate box. An atom of each element has one more electron and one more proton than an atom of the element to its left. Each horizontal row of the table is called a **period.** Changes in chemical properties of elements across a period correspond to changes in the electron arrangements of their atoms. Each vertical column of the table, known as a **group,** lists elements with similar properties. The elements in a group have similar chemical properties because their atoms have the same number of electrons in their outer energy level. For example, the elements helium, neon, argon, krypton, xenon, and radon all have similar properties and are known as the noble gases.

Molecules and Compounds

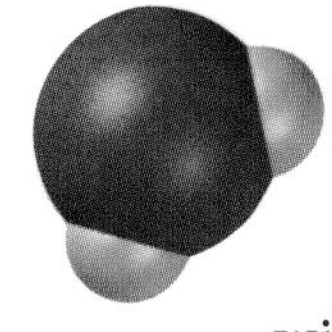

When two or more elements are joined chemically, the resulting substance is called a **compound.** A compound is a new substance with properties different from those of the elements that compose it. For example, water, H_2O, is a compound formed when hydrogen (H) and oxygen (O) combine. The smallest complete unit of a compound that has the properties of that compound is called a **molecule.** A chemical formula indicates the elements in a compound. It also indicates the relative number of atoms of each element present. The chemical formula for water is H_2O, which indicates that each water molecule consists of two atoms of hydrogen and one atom of oxygen. The subscript number is used after the symbol for an element to indicate how many atoms of that element are in a single molecule of the compound.

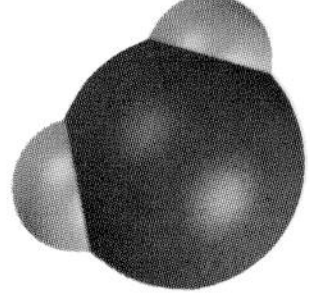

Acids, Bases, and pH

An ion is an atom or group of atoms that has an electric charge because it has lost or gained one or more electrons. When an acid, such as hydrochloric acid, HCl, is mixed with water, it separates into ions. An **acid** is a compound that produces hydrogen ions, H^+, in water. The hydrogen ions then combine with a water molecule to form a hydronium ion, H_3O^+. A **base,** on the other hand, is a substance that produces hydroxide ions, OH^-, in water.

To determine whether a solution is acidic or basic, scientists use pH. The **pH** is a measure of the hydronium ion concentration in a solution. The pH scale ranges from 0 to 14. The middle point, pH = 7, is neutral, neither acidic nor basic. Acids have a pH less than 7; bases have a pH greater than 7. The lower the number is, the more acidic the solution. The higher the number is, the more basic the solution.

Chemical Equations

A chemical reaction occurs when a chemical change takes place. (In a chemical change, new substances with new properties are formed.) A chemical equation is a useful way of describing a chemical reaction by means of chemical formulas. The equation indicates what substances react and what the products are. For example, when carbon and oxygen combine, they can form carbon dioxide. The equation for the reaction is as follows:

$$C + O_2 \rightarrow CO_2.$$

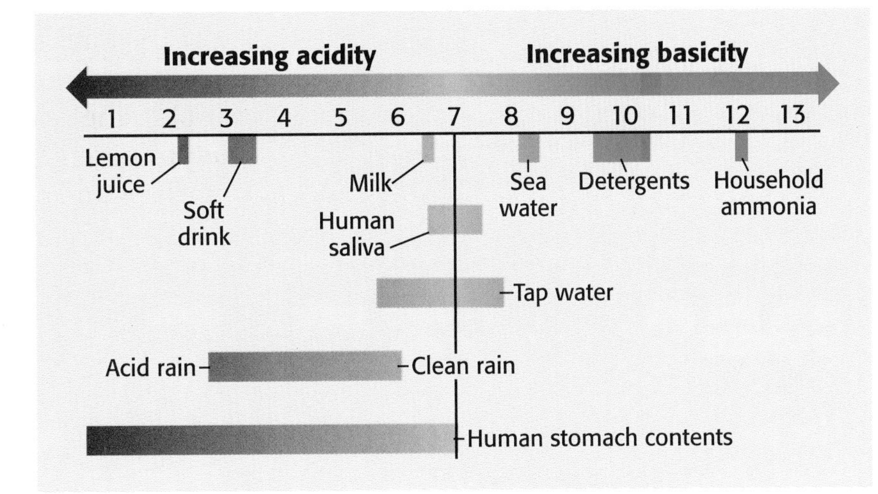

Properties of Common Minerals

	Mineral	Color	Luster	Streak	Hardness
Silicate Minerals	**Beryl**	deep green, pink, white, bluish green, or light yellow	vitreous	none	7.5–8
	Chlorite	green	vitreous to pearly	pale green	2–2.5
	Garnet	green or red	vitreous	none	6.5–7.5
	Hornblende	dark green, brown, or black	vitreous or silky	none	5–6
	Muscovite	colorless, gray, or brown	vitreous or pearly	white	2–2.5
	Olivine	olive green	vitreous	none	6.5–7
	Orthoclase	colorless, white, pink, or other colors	vitreous to pearly	white or none	6
	Plagioclase	blue gray to white	vitreous	white	6
	Quartz	colorless or white; any color when not pure	vitreous or waxy	white or none	7
Nonsilicate Minerals	**Native Elements**				
	Copper	copper-red	metallic	copper-red	2.5–3
	Diamond	pale yellow or colorless	vitreous	none	10
	Graphite	black to gray	submetallic	black	1–2
	Carbonates				
	Aragonite	colorless, white, or pale yellow	vitreous	white	3.5–4
	Calcite	colorless or white to tan	vitreous	white	3
	Halides				
	Fluorite	light green, yellow, purple, bluish green, or other colors	vitreous	none	4
	Halite	colorless or gray	vitreous	white	2.5–3
	Oxides				
	Hematite	reddish brown to black	metallic to earthy	red to red-brown	5.6–6.5
	Magnetite	iron black	metallic	black	5–6
	Sulfates				
	Anhydrite	colorless, bluish, or violet	vitreous to pearly	white	3–3.5
	Gypsum	white, pink, gray, or colorless	vitreous, pearly, or silky	white	1–2.5
	Sulfides				
	Galena	lead gray	metallic	lead gray to black	2.5
	Pyrite	brassy yellow	metallic	greenish, brownish, or black	6–6.5

APPENDIX

Density (g/cm^3)	Cleavage, Fracture, Special Properties	Common Uses
2.6–2.8	1 cleavage direction; irregular fracture; some varieties fluoresce in ultraviolet light	gemstones, ore of the metal beryllium
2.6–3.3	1 cleavage direction; irregular fracture	
4.2	no cleavage; conchoidal to splintery fracture	gemstones, abrasives
3.2	2 cleavage directions; hackly to splintery fracture	
2.7–3	1 cleavage direction; irregular fracture	electrical insulation, wallpaper, fireproofing material, lubricant
3.2–3.3	no cleavage; conchoidal fracture	gemstones, casting
2.6	2 cleavage directions; irregular fracture	porcelain
2.6–2.7	2 cleavage directions; irregular fracture	ceramics
2.6	no cleavage; conchoidal fracture	gemstones, concrete, glass, porcelain, sandpaper, lenses
8.9	no cleavage; hackly fracture	wiring, brass, bronze, coins
3.5	4 cleavage directions; irregular to conchoidal fracture	gemstones, drilling
2.3	1 cleavage direction; irregular fracture	pencils, paints, lubricants, batteries
2.95	2 cleavage directions; irregular fracture; reacts with hydrochloric acid	minor source of barium
2.7	3 cleavage directions; irregular fracture; reacts with weak acid, double refraction	cements, soil conditioner, whitewash, construction materials
3.2	4 cleavage directions; irregular fracture; some varieties fluoresce or double refract	hydrochloric acid, steel, glass, fiberglass, pottery, enamel
2.2	3 cleavage directions; splintery to conchoidal fracture; salty taste	tanning hides, fertilizer, salting icy roads, food preservation
5.25	no cleavage; splintery fracture; magnetic when heated	iron ore for steel, gemstones, pigments
5.2	2 cleavage directions; splintery fracture; magnetic	iron ore
2.89–2.98	3 cleavage directions; conchoidal to splintery fracture	soil conditioner, sulfuric acid
2.2–2.4	3 cleavage directions; conchoidal to splintery fracture	plaster of Paris, wallboard, soil conditioner
7.4–7.6	3 cleavage directions; irregular fracture	batteries, paints
5	no cleavage; conchoidal to splintery fracture	dyes, inks, gemstones

Glossary

A

absolute dating the process of establishing the age of an object, such as a fossil or rock layer, by determining the number of years it has existed (64)

asthenosphere (as THEN uh SFIR) the soft layer of the mantle on which pieces of the lithosphere move (90)

C

caldera (kahl DER uh) a circular depression that forms when a magma chamber empties and causes the ground above to sink (155)

cast an object created when sediment fills a mold and becomes rock (70)

catastrophism a principle that states that all geologic change occurs suddenly (57)

cinder cone volcano a small, steeply sloped volcano that forms from moderately explosive eruptions of pyroclastic material (154)

cleavage (KLEEV IJ) the tendency of a mineral to break along flat surfaces (9)

composite volcano a volcano made of alternating layers of lava and pyroclastic material; also called *stratovolcano* (154)

composition the makeup of a rock; describes either the minerals or elements present in it (31)

compound a pure substance made of two or more elements that have been chemically joined, or bonded together (5)

compression the type of stress that occurs when an object is squeezed (103)

continental drift the theory that continents can drift apart from one another and have done so in the past (95)

convergent boundary the boundary between two colliding tectonic plates (100, 159)

coprolites (KAHP roh LIETS) preserved feces, or dung, from animals (70)

core the central, spherical part of the Earth below the mantle (89)

crater a funnel-shaped pit around the central vent of a volcano (155)

crust the thin, outermost layer of the Earth, or the uppermost part of the lithosphere (88, 159)

crystal the solid, geometric form of a mineral produced by a repeating pattern of atoms (5)

D

deformation the change in the shape of rock in response to stress (103, 121)

density the amount of matter in a given space; mass per unit volume (10)

divergent boundary the boundary between two tectonic plates that are moving away from each other (101)

E

elastic rebound the sudden return of elastically deformed rock to its original shape (121)

element a pure substance that cannot be separated or broken down into simpler substances by ordinary chemical means (4)

eon the largest division of geologic time (75)

epicenter the point on the Earth's surface directly above an earthquake's starting point (126)

epoch (EP uhk) the fourth-largest division of geologic time (75)

era the second-largest division of geologic time (75)

erosion the removal and transport of material by wind, water, or ice (28–29, 62)

extrusive (eks TROO siv) the type of igneous rock that forms when lava or pyroclastic material cools and solidifies on the Earth's surface (36)

F

fault a break in the Earth's crust along which blocks of the crust slide relative to one another due to tectonic forces (105, 120)

fault block a block of the Earth's crust on one side of a fault (105)

focus the point inside the Earth where an earthquake begins (126)

folding the bending of rock layers due to stress in the Earth's crust (104)

foliated the texture of metamorphic rock in which the mineral grains are aligned like the pages of a book (44)

footwall the fault block that is below a fault (105)

fossil any naturally preserved evidence of life (68)

fracture the tendency of a mineral to break along curved or irregular surfaces (9)

G

gap hypothesis states that sections of active faults that have had relatively few earthquakes are likely to be the sites of strong earthquakes in the future (131)

geologic column an ideal sequence of rock layers that contains all the known fossils and rock formations on Earth arranged from oldest to youngest (60)

geologic time scale a scale that divides Earth's 4.6-billion-year history into distinct intervals of time (74)

geothermal energy energy from within the Earth (116)

H

half-life for a particular radioactive sample, the time it takes for one-half of the sample to decay (65)

hanging wall the fault block that is above a fault (105)

hardness the resistance of a mineral to being scratched (10)

hot spot a place on Earth's surface that is directly above a column of rising magma called a mantle plume (159)

I

igneous rock rock that forms from the cooling of magma (30)

index fossil a fossil of an organism that lived during a relatively short, well-defined time span; a fossil that is used to date the rock layers in which it is found (72)

inner core the solid, dense center of the Earth (91)

intrusive (in TROO siv) the type of igneous rock that forms when magma cools and solidifies beneath Earth's surface (35)

isotopes atoms of the same element that have the same number of protons but have different numbers of neutrons (64)

L

lava magma that flows onto the Earth's surface (30, 148)

lithosphere (LITH oh SFIR) the outermost, rigid layer of the Earth that consists of the crust and the rigid upper part of the mantle (90)

luster the way the surface of a mineral reflects light (8)

M

magma the hot liquid that forms when rock partially or completely melts; may include mineral crystals (29)

magnetic reversal the process by which the Earth's north and south magnetic poles periodically change places (98)

mantle the layer of the Earth between the crust and the core (89, 156)

mesosphere literally, the "middle sphere"—the strong, lower part of the mantle between the asthenosphere and the outer core (91)

metamorphic rock rock that forms when the texture and composition of preexisting rock changes due to heat or pressure (30)

meteorite a meteoroid that reaches the Earth's surface without burning up completely (27)

mid-ocean ridge a long mountain chain that forms on the ocean floor where tectonic plates pull apart; usually extends along the center of ocean basins (97, 101, 158)

mineral a naturally formed, inorganic solid with a crystalline structure (4)

Moho a place within the Earth where the speed of seismic waves increases sharply; marks the boundary between the Earth's crust and mantle (135)

mold a cavity in the ground or rock where a plant or animal was buried (70)

N

natural gas a gaseous fossil fuel (106)

nonfoliated the texture of metamorphic rock in which mineral grains show no alignment (44)

nonrenewable resource a natural resource that cannot be replaced or that can be replaced only over thousands or millions of years (15)

nonsilicate mineral a mineral that does not contain compounds of silicon and oxygen (7)

normal fault a fault in which the hanging wall moves down relative to the footwall (105)

O

ore a mineral deposit large enough and pure enough to be mined for a profit (14)

outer core the liquid layer of the Earth's core that lies beneath the mantle and surrounds the inner core (91)

P

period the third-largest division of geologic time (75)

permineralization a process in which minerals fill in pore spaces of an organism's tissues (68)

petrification a process in which an organism's tissues are completely replaced by minerals (68)

plate tectonics the theory that the Earth's lithosphere is divided into tectonic plates that move around on top of the asthenosphere (99)

P waves the fastest type of seismic wave; can travel through solids, liquids, and gases; also known as pressure waves and primary waves (124)

pyroclastic material fragments of rock that are created by explosive volcanic eruptions (151)

R

radioactive decay a process in which radioactive isotopes tend to break down into stable isotopes of other elements (64)

radiometric dating determining the absolute age of a sample based on the ratio of parent material to daughter material (65)

reclamation the process of returning land to its original state after mining is completed (15)

relative dating determining whether an object or event is older or younger than other objects or events (59)

reverse fault a fault in which the hanging wall moves up relative to the footwall (105)

rift a deep crack that forms between tectonic plates as they separate (158)

rock a solid mixture of crystals of one or more minerals or other materials (26)

rock cycle the process by which one rock type changes into another rock type (28)

S

sea-floor spreading the process by which new oceanic lithosphere is created at mid-ocean ridges as older materials are pulled away from the ridge (97)

sedimentary rock rock that forms when sediments are compacted and cemented together (30)

seismic (SIEZ mik) **gap** an area along a fault where relatively few earthquakes have occurred (131)

seismic waves waves of energy that travel through the Earth (124)

seismogram a tracing of earthquake motion created by a seismograph (126)

seismograph an instrument located at or near the surface of the Earth that records seismic waves (126)

seismology the study of earthquakes (120)

shadow zone an area on the Earth's surface where no direct seismic waves from a particular earthquake can be detected (135)

shield volcano a large, gently sloped volcano that forms from repeated, nonexplosive eruptions of lava (154)

silica a compound of silicon and oxygen atoms (150)

silicate mineral a mineral that contains a combination of the elements silicon and oxygen (6)

strata layers of sedimentary rock that form from the deposition of sediment (37)

stratification the layering of sedimentary rock (40)

streak the color of a mineral in powdered form (9)

stress the amount of force per unit area that is put on a given material (103)

strike-slip fault a fault in which the two fault blocks move past each other horizontally (106)

subduction zone the region where an oceanic plate sinks down into the asthenosphere at a convergent boundary, usually between continental and oceanic plates (159)

superposition a principle that states that younger rocks lie above older rocks in undisturbed sequences (59)

S waves the second-fastest type of seismic wave; cannot travel through materials that are completely liquid; also known as shear waves and secondary waves (124)

T

tectonic plate a piece of the lithosphere that moves around on top of the asthenosphere (92)

tension the type of stress that occurs when forces act to stretch an object (103)

texture the sizes, shapes, and positions of the grains that a rock is made of (32)

trace fossil any naturally preserved evidence of an animal's activity (70)

transform boundary the boundary between two tectonic plates that are sliding past each other horizontally (101)

U

unconformity a surface that represents a missing part of the geologic column (62)

uniformitarianism a principle that states that the same geologic processes shaping the Earth today have been at work throughout Earth's history (56)

V

volcano a mountain that forms when molten rock, called magma, is forced to the Earth's surface (148)

Index

A **boldface** number refers to an illustration on that page.

N

O

P

Q

R

S

T

U

V

W

Credits

Abbreviations used: (t) top, (c) center, (b) bottom, (l) left, (r) right, (bkgd) background

ILLUSTRATIONS

All work, unless otherwise noted, contributed by Holt, Rinehart & Winston.

Table of Contents: iv(bl), Uhl Studios, Inc.; v(tr), Uhl Studios, Inc.

Chapter One: Page 4(bl), Gary Locke; 5, Stephen Durke/Washington Artists; 12-13(bkgd), Uhl Studios, Inc.; 14(bl), Jared Schneidman Design.

Chapter Two: Page 28-29, Uhl Studios, Inc.; 31(b), Sidney Jablonski; 33, Keith Locke; 34(l), Uhl Studios, Inc.; 35(b), Uhl Studios, Inc.; 38(bl), Robert Hynes; 42(b), Uhl Studios, Inc.; 43(t), Stephen Durke/Washington Artists; 43(b), Uhl Studios, Inc.; 48(br), Sidney Jablonski; 51(cr), Sidney Jablonski.

Chapter Three: Page 56(b), Uhl Studios, Inc.; 58(c), Barbara Hoopes-Ambler; 60(b), Jared Schneidman Design; 61, Uhl Studios, Inc.; 62(b), Jared Schneidman Design; 63, Uhl Studios, Inc.; 64(b), Stephen Durke/Washington-Artists' Represents; 69(br), Will Nelson/Sweet Reps; 70-71(c), Frank Ordaz; 72(c), Uhl Studios, Inc.; 83(tr), Joe LeMonnier.

Chapter Four: Page 88(b), Uhl Studios, Inc.; 89(br), Uhl Studios, Inc.; 90-91, Uhl Studios, Inc.; 92(c), Uhl Studios, Inc.; 93(c), Uhl Studios, Inc.; 94(tl), Uhl Studios, Inc.; 95(tr), Uhl Studios, Inc.; 95(bl), MapQuest.com; 96, MapQuest.com; 97, Uhl Studios, Inc.; 98(cl), Stephen Durke/Washington Artists; 98(cr), Uhl Studios, Inc.; 99(b), Uhl Studios, Inc.; 100-101(b), Uhl Studios, Inc.; 104(tl), Uhl Studios, Inc.; 105(tr), Marty Roper/Planet Rep; 105(cr), Uhl Studios, Inc.; 105(br), Uhl Studios, Inc.; 107(tr), Uhl Studios, Inc.; 108(t), Tony Morse/Ivy Glick; 108(bl), Uhl Studios, Inc.; 112, Uhl Studios, Inc.; 113(cr), Marty Roper/Planet Rep.

Chapter Five: Page 120(b), MapQuest.com; 121(b), Uhl Studios, Inc.; 122(b), Uhl Studios, Inc.; 122-123, Uhl Studios, Inc.; 124, Uhl Studios, Inc.; 125(cl), Uhl Studios, Inc.; 126(bl), Uhl Studios, Inc.; 127(tr), Sidney Jablonski; 129(b), MapQuest.com; 131(t), Jared Schneidman Design; 132, Uhl Studios, Inc.; 135(b), Uhl Studios, Inc.; 136, Sidney Jablonski; 140(br), Sidney Jablonski; 140(cl), Uhl Studios, Inc.; 142(br), Uhl Studios, Inc.; 143(cr), Sidney Jablonski.

Chapter Six: Page 150(tl), Uhl Studios, Inc.; 154(l), Patrick Gnan; 155(tr), Uhl Studios, Inc.; 156(b), Uhl Studios, Inc.; 157(tr), Stephen Durke/Washington Artists; 157(bl), MapQuest.com; 158, Uhl Studios, Inc.; 159, Uhl Studios, Inc.; 160(t), Uhl Studios, Inc.; 167(tr), Ross, Culbert and Lavery.

LabBook: Page 175(tr), Mark Heine; 176(br), Mark Heine; 182, Mark Heine; 184(c), Sidney Jablonski; 185(c), MapQuest.com; 186(t), Marty Roper/Planet Rep; 187(tr), Ralph Garafola/Lorraine Garafola Represents.

Appendix: Page 190(c), Terry Guyer; 194(b), Mark Mille/Sharon Langley; 202, Kristy Sprott; 203, Kristy Sprott; 204(bl), Stephen Durke/Washington Artists; 205(tl), Stephen Durke/Washington Artists; 205(c), Stephen Durke/Washington Artists; 205(b), Bruce Burdick.

PHOTOGRAPHY

Cover and Title Page: Frans Lanting/Minden Pictures

Sam Dudgeon/HRW Photo: Page viii-1, 4, 6(cr), 9(tr,bl), 10(bl), 11(tc,cl), 12(c), 13(b), 19(l), 20(tl), 31(c), 32(cl), 38(c), 41(cl), 57(all), 59(bl), 65, 103, 133(tr), 134(br), 170, 171(bc), 172(br,cl), 173(tl, b), 177(all), 178, 179, 181, 183, 191(br).

Table of Contents: iv(tl), E. R. Degginger/Color-Pic, Inc.; v(cr), Stephen Frink/Corbis; v(b), Robert Glusic/Natural Selection; vi(tl), E. R. Degginger/Color-Pic, Inc.; vi(bl), Sam Dudgeon/HRW Photo; vii(tr), A. F. Kersting; vii(cr), SuperStock; vii(br), Alberto Garcia/SABA.

Chapter One: pp. 2-3 Luis Rosendo/FPG International; p. 3 HRW Photo; p. 3 Inga Spence/Tom Stack & Associates; p. 5(br), Dr. Rainer Bode/Bode-Verlag Gmb; p. 6(bl), Pat Lanza/Bruce Coleman Inc.; p. 6(cl, c), E. R. Degginger/Color-Pic, Inc.; p. 7(top to bottom), (top four), E. R. Degginger/Color-Pic, Inc.; SuperStock; Visuals Unlimited/Ken Lucas; p. 8(tr), Liaison Agency; (tl), Jane Burton/Bruce Coleman, Inc.; Luster Chart (row 1), E. R. Degginger/Color-Pic, Inc.; John Cancalosi 1989/DRK Photo; (row 2), Biophoto Associates/Photo Researchers, Inc.; Dr. E. R. Degginger/Bruce Coleman Inc.; (row 3), E. R. Degginger/Color-Pic, Inc.; Biophoto Associates/Photo Researchers, Inc.; (row 4), E. R. Degginger/Color-Pic, Inc.; p. 9(br), Tom Pantages; p. 9(c), E. R. Degginger/Color-Pic, Inc.; p. 9(cl), Erica and Harold Van Pelt/American Museum of Natural History; p. 10(0), Visuals Unlimited/Ken Lucas; 10(0,0,0,0), E. R. Degginger/Color-Pic, Inc.; p. 10(0), Visuals Unlimited/Dane S. Johnson; p. 10(0), Carlyn Iverson/Absolute Science Illustration and Photography; p. 10(0), Mark A. Schneider/Visuals Unlimited; p. 10(0), Charles D. Winters/Photo Researchers, Inc.; p. 10(00), Bard Wrisley/Liasion Agency; p. 11(tl, bc), E. R. Degginger/Color-Pic, Inc.; p. 11(tr), Sam Dudgeon/HRW Photo Courtesy Science Stuff, Austin , TX; p. 11(cr), Tom Pantages Photography; p. 11(br), Victoria Smith/HRW Photo; p. 12(b), E. R. Degginger/Color-Pic, Inc.; p. 12(tl), Victoria Smith/HRW Photo, Courtesy Science Stuff, Austin, TX; p. 13(cr, t), E. R. Degginger/Color-Pic, Inc.; p. 14(c), Wernher Krutein/Liaison Agency; p. 14 (inset), Kosmatsu Mining Systems; p. 14, Index Stock Photography, Inc.; p. 15, Historic Royal Palaces; p. 17 tr Victoria Smith/Courtesy of Science Stuff, Austin, TX/HRW Photo; p. 17 tcr Ken Lucas/Visuals Unlimited; p. 17 cr Charile Winters/HRW Photo; p. 17 (bcr, br) Sam Dudgeon/HRW Photo; p. 19(inset), Kosmatsu Mining Systems; p. 19(tr), Wernher Krutein/Liaison Agency; p. 20(bl), E. R. Degginger/Color-Pic, Inc.; p. 22(bl), Peter Menzel; p. 22(bl), Ralph Wetmore/Stone.

Chapter Two: pp. 24-25 Tom Till; p. 25 HRW Photo; p. 26(c), Kenneth Garrett; p. 26(bc), Fergus O'Brian/FPG International; p. 26(br), Peter Cummings/Tom Stack & Associates; p. 26(bl), Historical Collections, National Museum of Health and Medicine, AFIP; p. 27(bl), A.F. Kersting; p. 27(br), Andy Christiansen/HRW Photo; p. 27(c), Breck P. Kent; p. 27(cr), NASA/Science Photo Library/Photo Researchers, Inc.; p. 31(granite), Pat Lanza/Bruce Coleman Inc.; p. 31(br,cr), E. R. Degginger/Color-Pic, Inc.; p. 31(cl), Walter H. Hodge/Peter Arnold; p. 31(bl), Sp. Harry Taylor/Dorling Kindersley; p. 31(bc), Breck P. Kent; p. 32(c), Dorling Kindersley; p. 32(cr bc), Breck P. Kent; p. 34(br), E. R. Degginger/Color-Pic, Inc.; p. 34(cl, bl, cr), Breck P. Kent; p. 35(tr), Laurence Parent; p. 36(cl), Breck P. Kent; p. 36(cr), Peter Frenck/Bruce Coleman, Inc.; p. 37(br), Robert Glusic/Natural Selection; p. 38(tl), Breck P. Kent/Animals Animals/Earth Scenes; 38(breccia), Breck P. Kent; p. 38(cl), Joyce Photographics/Photo Researchers, Inc.; p. 38(cr), E. R. Degginger/Color-Pic, Inc.; p. 38(br), Breck P. Kent; p. 39(br), Ed Cooper; p. 39, Stephen Frink/Corbis; p. 39(bl), Breck P. Kent; p. 39(c), SuperStock; p. 40(cr), Franklin P. OSF/Animals Animals/Earth Scenes; p. 40(tl), Breck P. Kent; p. 41(bl), E. R. Degginger/Color-Pic, Inc.; p. 41(br), George Wuerthner; p. 43(tl), Visuals Unlimited/Dane S. Johnson; p. 43(tlc), Carlyn Iverson/Absolute Science Illustration and Photography; p. 43(tlb), Breck P. Kent; p. 43(tr), Breck P. Kent/Animals Animals/Earth Scenes; p. 43(brc), Tom Pantages; p. 43(br), Breck P. Kent/Animals Animals/Earth Scenes; p. 44(tl), Ken Karp/HRW Photo; p. 44(br), Breck P. Kent; p. 45(tl), E. R. Degginger/Color-Pic, Inc.; p. 45(bl), Ray Simmons/ Photo Researchers, Inc.; p. 45(tc), The Natural History Museum, London; p. 45(bc), Breck P. Kent; p. 46 Peter Van Steen//HRW Photo; p. 47 Peter Van Steen//HRW Photo; p. 48, E. R. Degginger/Color-Pic, Inc.; p. 49, Doug Sokell/Tom Stack & Associates; p. 52(c), Wolfgang Kaehler/Liason International.

Chapter Three: pp. 54-55 Blair Jonathan/National Geographic Society Image Collection; p. 54 Brett Gregory/Auscape International Pty Ltd.; p. 55 HRW Photo; p. 59(br), Andy Christiansen/HRW Photo; p. 66, Tom Till/DRK Photo; p. 67, Courtesy Charles S. Tucek/University of Arizona at Tucson; p. 68(bl), Francois Gohier/Photo Researchers, Inc.; p. 68(cr), p. 147(tr), E. R. Degginger/Color-Pic, Inc.; p. 70(cl), Breck P. Kent; p. 70(bl), The G.R. "Dick" Roberts Photo Library; p. 71(c, cr), Brian Exton; p. 72(tr), Thomas R. Taylor/Photo Researchers, Inc.; p. 73(br), Mike Buchheit Photography; p. 75, p. 76 (all), American Museum of Natural History; p. 82(tl), Runk/Schoenberger/Grant Heilman Photography; p. 82(cr), Stone; p. 84(c), Andrew Leitch/©1992 The Walt Disney Co. Reprinted with the permission of Discover Magazine; p. 85(br, tl), Louie Psihoyos/Matrix.

Chapter Four: pp. 86-87 Jock Montgomery/Bruce Coleman, Inc.; p. 86 Courtesy of the Fort Wroth Museum of Science and History; p. 87 HRW Photo; p. 89(tr), James Wall/Animals Animals/Earth Scenes; p. 89(bl), World Perspective/Stone; p. 102(cl), ESA/CE/Eurocontrol/Science Photo Library/Photo Researchers, Inc.; p. 102(tl), NASA; p. 104(br), Visuals Unlimited/SylvesterAllred; p. 104(bl), p. 184(br), The G.R. "Dick" Roberts Photo Library; 106(tl), Tom Bean; p. 106(tr), Landform Slides; p. 107(bl), William Manning/The Stock Market; p. 109, Michelle & Tom Grimm/Stone; p. 110 Sam Dudgeon/HRW Photo; p. 111 Sam Dudgeon/HRW Photo; p. 114, NASA/Photo Researchers, Inc.; p. 116, Bob Krist; p. 117, Martin Schwarzbach/Photo Deutsches Museum Munchen.

Chapter Five: pp. 118-119 Wally Santana/AP/Wide World Photo; p. 118 Haley/SIPA Press; p. 119 HRW Photo; p. 121, Joe Dellinger/NOAA/National Geophysical Data Center; p. 126(cl), Bob Paz/Caltech; p. 127, Earth Images/Stone; p. 130, Peter Cade/Stone; p. 131, A. Ramey/Woodfin Camp & Associates; p. 133(bl), Paul Chesley/Stone; p. 134(tl), Ken Lax; p. 137(cr), NASA; p. 137(tr), SOHO (ESA & NASA); p. 139 Sam Dudgeon/HRW Photo; p. 141, A. Ramey/Woodfin Camp & Associates; p. 142(tl), Chuck O'Rear/Corbis; p. 144(c), Novaswan/FPG International; p. 145(cr), David Madison/Bruce Coleman, Inc.

Chapter Six: pp. 146-147 Carl Shaneff/Pacific Stock; p. 146 Darodents/Pacific Stock; p. 147 HRW Photo; p. 148(cl), Robert W. Madden/National Geographic Society; p. 148(cr), Douglas Peebles Photography; 148(bc), Ken Sakamoto/Black Star; p. 149(b), Breck P. Kent/Earth Scenes; p. 149(tr), Joyce Warren/USGS Photo Library; p. 151(cr), Jim Yuskavitch; p. 151(bl), Karl Weatherly; p. 151(bc), Tui De Roy/Minden Pictures; p. 151(br), B. Murton/

continued on page 216

Self-Check Answers

Chapter 1—Minerals of the Earth's Crust

Page 13: These minerals form wherever salt water has evaporated.

Chapter 2—Rocks: Mineral Mixtures

Page 34: From fastest-cooled to slowest-cooled, the rocks in Figure 10 are: basalt, rhyolite, gabbro, and granite.

Page 42: A rock can come into contact with magma and also be subjected to pressure underground.

Chapter 3—The Rock and Fossil Record

Page 70: Coprolites and tracks are trace fossils because they are evidence of animal activity rather than fossilized organisms.

Chapter 4—Plate Tectonics

Page 105: When folding occurs, sedimentary rock strata bend but do not break. When faulting occurs, sedimentary rock strata break along a fault, and the fault blocks on either side move relative to each other.

Chapter 5—Earthquakes

Page 123: Convergent motion creates reverse faults, while divergent motion creates normal faults. Convergent motion produces deep, strong earthquakes, while divergent motion produces shallow, weak earthquakes.

Page 130: 120

Chapter 6—Volcanoes

Page 157: Solid rock may become magma when pressure is released, when the temperature rises above its melting point, or when its composition changes.

Credits ***(continued)***

Southampton Oceanography Centre/Science Photo Library/Photo Researchers, Inc.; p. 152(tl), Tom Bean/DRK Photo; p. 152(t), Francois Gohier/Photo Researchers, Inc.; p. 152(c), Visuals Unlimited/Glenn Oliver; p. 152(b), E. R. Degginger/Color-Pic, Inc.; p. 153, Alberto Garcia/SABA; p. 154(tl), Visuals Unlimited/Jeff Greenberg; p. 154(cl), Krafft/Explorer/Science Source/Photo Researchers, Inc.; p. 154(bl), SuperStock; p. 160(bl), Andrew Rafkind/Stone; p. 161(tr), Game McGimsey/ USGS Alaska Volcano Observatory; p. 161(br), Gilles Bassignac/Liaison Agency; p. 164(c), Robert W. Madden/National Geographic Society; p. 166(tl), Krafft/ Explorer/Science Source/Photo Researchers, Inc.; 166(bl), Karl Weatherly; p. 168(c), The Robotics Institute Carnegie Mellon University; p. 169(bl), NASA/ Science Photo Library/Photo Researchers, Inc.

LabBook/Appendix: "LabBook Header", "L", Corbis Images; "a", Letraset Phototone; "b", and "B", HRW; "o", and "k", images ©2001 PhotoDisc/HRW; Page 171(tr), John Langford/HRW Photo; 171(cl), Michelle Bridwell/HRW Photo; 171(br), Image ©2001 PhotoDisc, Inc./HRW; 172(bl), Stephanie Morris/ HRW Photo; 173(b) Peter Van Steen/HRW Photo; 173(tl) Sam Dudgeon/HRW Photo; 173(tr), Jana Birchum/HRW Photo; 174, Victoria Smith/HRW Photo; 177(bc), Russell Dian/HRW Photo; 180, Andy Christiansen/HRW Photo; 183(b), James Tallon/Outdoor Exposures; 184, Andy Christiansen/HRW Photo; 191(tr), Peter Van Steen/HRW Photo.

Feature Borders: Unless otherwise noted below, all images ©2001 PhotoDisc/HRW. "Across the Sciences" 169, all images by HRW; "Careers" 85, sand bkgd and Saturn, Corbis Images; DNA, Morgan Cain & Associates; scuba gear, ©1997 Radlund & Associates for Artville; "Eye on the Environment" 145, clouds and sea in bkgd, HRW; bkgd grass, red eyed frog, Corbis Images; hawks, pelican, Animals Animals/Earth Scenes; rat, Visuals Unlimited/John Grelach; endangered flower, Dan Suzio/Photo Researchers, Inc.; "Health Watch" 53, dumbbell, Sam Dudgeon/HRW Photo; aloe vera, EKG, Victoria Smith/HRW Photo; basketball, ©1997 Radlund & Associates for Artville; shoes, bubbles, Greg Geisler; "Scientific Debate" 117, Sam Dudgeon/HRW Photo; "Science Fiction" 23, saucers, Ian Christopher/Greg Geisler; book, HRW; bkgd, Stock Illustration Source; "Science Technology and Society" 52, 84, 116, 168, robot, Greg Geisler; "Weird Science" 22, 144, mite, David Burder/Tony Stone; atom balls, J/B Woolsey Associates; walking stick, turtle, EclectiCollection.